RIGGING

RIGGING

Everything you always wanted to know about the ropes and the rigging, the winches and the mast of a cruising or racing boat

Danilo Fabbroni

Translated by Martyn Drayton

Authorised translation from the Italian language edition published by Editrice Incontri Nautici, April 2007

Email (for orders and customer service enquiries): cs-books@wiley.co.uk
Visit our Home Page on www.wiley.com

Other Wiley Editorial Offices

John Wiley & Sons Inc., 111 River Street, Hoboken, NJ 07030, USA

Jossey-Bass, 989 Market Street, San Francisco, CA 94103-1741, USA

Wiley-VCH Verlag GmbH, Boschstr. 12, D-69469 Weinheim, Germany

John Wiley & Sons Australia Ltd, 42 McDougall Street, Milton, Queensland 4064, Australia

John Wiley & Sons (Asia) Pte Ltd, 2 Clementi Loop #02-01, Jin Xing Distripark, Singapore 129809

John Wiley & Sons Canada Ltd, 6045 Freemont Blvd. Mississauga, Ontario, L5R 4J3 Canada

Wiley also publishes its books in a variety of electronic formats. Some content that appears in print may not be available in electronic books.

Library of Congress Cataloging-in-Publication Data

Fabbroni, Danilo.
Rigging / Danilo Fabbroni ; translated by Martyn Drayton.
p. cm.
Translation from the Italian edition, also entitled Rigging.
ISBN 978-0-470-72568-9 (pbk. : alk. paper)
1. Masts and rigging. I. Title.
VM531.F33 2008
623.8′62 – dc22

2007050171

British Library Cataloguing in Publication Data

A catalogue record for this book is available from the British Library

ISBN: 978-0-470-72568-9

Typeset 10/12pt Futura by SNP Best-set Typesetter Ltd., Hong Kong

Printed and bound by TJ International Ltd, Padstow, Cornwall

Contents

Preface vii

Acknowledgements ix

1 Running rigging 1

2 Genoa sheets: the forces involved 47

3 The genoa cars 70

4 The mainsheet 85

5 Spinnaker sheets and afterguys 109

6 Halyards and reef lines 114

7 Standing rigging 122

8 Setting up a swept back rig 149

9 Winches 155

Bibliography 171

To Angela and Erica

Preface

In yachting – that is, sailing the seas for pleasure, as opposed to doing so for profit, as in the merchant marine, or for defence, as in the navy – rigging is just a tile in the overall mosaic of sailing.

Only a tile, but a vitally important one. A tile that literally unites the hull, mast and sails into a single entity, and that allows the crew to optimise the boat's performance according to the needs of the moment.

Besides, to use a parallel from the automobile sector, what would an engine, even the best one of its kind, be worth without a suitable transmission system?

Well, rigging is simply the transmission system whose job is to transmit to the hull the power harnessed by the sails. This brings to mind a significant episode that demonstrates the meaning of rigging.

At the end of the summer of 1985 I had to sail *Brava Les Copains*, a racing yacht, from Porto Cervo to Palma de Majorca, where the world One Ton Cup was to be held. During the trip, the belt of the alternator broke. I fixed this serious problem, which would soon have left us without instrumentation because of the lack of power, by splicing a length of Kevlar line into a strop to serve as a new belt. It lasted until we arrived in port.

The sense, the function and the aim of rigging is to transmit the energy potential of the sails to the hull. It does this with standing rigging, which holds the mast in place, and with running rigging, which is used to hoist and trim the sails.

But the finest, most precise and striking definition of rigging I know was given by an American rigger, Brion Toss:

'Rigging is the art of moving things or keeping them still with cords, pulleys and knots.'

The terms 'cords' and 'pulleys' were used intentionally, because contrary to the – what shall we call it – 'blinkered' vision of the sailor, rigging doesn't just mean the halyards, sheets, stays, blocks and winches of a sailing boat, but is a very broad field that goes from the cables of the huge dockyard cranes to the tiny block and tackle used to open the umbrella on our terrace; the myriad tie-rods that support such colossal bridges as the one in Denmark and the four lines on which we hang out our washing; the numberless steel rods that form the framework of the glass pyramid by I.M. Pei that stands in front of the Louvre in Paris, or the special recovery line used by alpine rescue helicopters.

We need the same breadth of vision when we speak of a *rig*.

A 'rig' does not only mean the way a sailing boat is fitted out, but also the framework that supports a theatre or concert hall[4] stage, and also the scaffolding used for maintenance work on buildings.

And now, finally, the point of a book on rigging is clear to me. It has become clear thanks to two considerations I have only been able to make since working for Harken. I have brought into focus what it is that the sailor, whether he be the helmsman of an Optimist or the owner of a Maxi, an incurable world cruiser or a fanatical racing man, wants to know, understand and master, taking it for granted that he's already able to handle a boat: is his boat fitted out in the best possible way for the job she's called upon to do?

This book aims to give clear answers to the dozens and dozens of questions a sailor asks himself, and asks technical people, but without getting – to judge from the frequency and the intensity of the questions – satisfactory answers.

If I have found and written down the answers to the present and future questions from the sailors I usually meet at the stands of boat shows, on the quayside and on the telephone in the company where I work, my task will be accomplished and I will have attained my objective. I will let the reader be the judge.

One final note. This is not, and does not aim to be, a book of knots and splices. I have always maintained that even a member of the Alpine Regiment, if well taught, could become a magnificent splicer. But splicing is not what it's about. Bernard Moitessier said that life is too short to learn how to splice! Splices are useful parts of the subject of rigging, but knowing how to do splices doesn't automatically make you a rigger or help you understand when and why to prefer one piece of rigging over another.

If the reader understands, from what I have written, this final 'why', he certainly won't have become a rigger, but he will undoubtedly be a more competent sailor. And I will have reached my goal.

Acknowledgements

As Brion Toss said, 'The hardest part of a book on rigging is the acknowledgements, which never end . . . because you have an infinite number of people to thank!'

For my part, first I must – and I really wish to – thank my parents, who in far-off 1970 literally shoved me – a boy born on the banks of an Alpine lake in Switzerland and raised on the shores of an Umbrian one, Lake Trasimeno – on board a Soling that, as soon as she started planing, infected me with an incurable disease: the passion for sailing! If my parents hadn't forced me on to that boat, all I could tell you about now would be how we played table football in the gardens by Lake Trasimeno!

Thanks to Admiral Di Giovanni, who appointed me seaman on board *Sagittario* during my military service in the navy: I learned from him much more than just how to go to sea.

I must also thank Sergio Doni, the unforgettable owner of *Yena*, which sailed the Sardinia Cup in 1980. He showed faith in a person like me, a freshwater sailor, banking just on my potential despite my lack of experience on big boats. Today, in a world dominated by the 'everything and now' philosophy, such an act would be rare indeed.

Thanks too to Pinin, Tacun and Chicco, 'historic' seamen whose skill equalled their generosity. But the 'tree' that bore them flowers no longer: the wood is finished. They weren't rare. They were unique.

Thanks also to people from outside Italy: first of all Ben Bradley of Spencer Rigging who generously welcomed me into his home . . . in remote times when I was undeniably homeless; thanks too to Curley, an unparalleled rigger at Spencer who, together with Pinin, taught me how to splice . . .

Thanks to Peter Morton of the English Riggarna, for his precious suggestions in the past.

Thanks to Eep Looman, the flying Dutchman of Illbruck's *Pinta* during the 1980s, who gave a warm welcome to the only Italian in a crew of Germans . . . in the middle of the Black Forest!

Thanks to Graham Fleury of Southern Ocean Ropes, and to Pierangelo Maffioli of Gottifredi & Maffioli for their invaluable advice on cordage.

Thanks to Lou Varney, an English friend with whom I've had many sailing adventures!

And thanks – a thousand thanks! – to Vittorio Vongher, whom I am still honoured to have as a friend, for having shared with me, for ten long years, an unforgettable, indescribable adventure: our rigging company Fabbroni & Vongher srl.[1] Without it this book could never have been written, though unfortunately circumstances prevented a direct critical dialogue during the writing.

Thanks to Giorgio Casti and Luigi Ciccarone, who were the first to urge that these pages should become a reality above and beyond the heap of drafts I had collected, and thanks to David Palmer who, on behalf of my English publisher, believed in what I had written.

Many thanks also to Giancarlo Basile and Enea Riboldi of *Bolina* for their invaluable reading and revision of the text.

Many thanks to Martyn Drayton, my able translator, who helped simplify the text and make it more effective.

Thanks to everyone in Harken and in particular to Alberto Lozza and Andrea Merello of the technical department, and to Luciano Bonassi,[2] for what I learned from them about winches, and to Giampaolo Spera, without whom the experience would have remained a daydream.

And lastly, but top of the list, heartfelt thanks to my wife Angela and to Erica: without their loving support I would never have found the strength to finish the task.

[1] The company is still active today in Porto S. Stefano, Grosseto.

[2] Luciano Bonassi, whom we should never stop mourning, was the *deus ex machina* of *Barbarossa*: a profound innovator, but with respect for tradition, in the field of winches, blocks and coffee grinder systems.

Running rigging

Runners – Why have Runners?

Runners at Work

Runner Blocks and Their Circuits

The Various Systems for Tensioning the Runners

The Runner Tail

Lower Runners

Mast Attachments of Upper and Lower Runners

The Backstay

How to choose wire rope, identify the right purchase and select the correct winch.

Why have Runners?

'Why have runners?' 'What use are they?' we may ask ourselves. Good questions! For without runners, the lives of sailors would be much easier! In fact, with the growing number of boats with aft raked spreaders,[1] the use of runners and the need for them seem to be dying out. I say 'seem to' because there is still an impressive number of boats around that use – and will continue to use – runners, both in racing and cruising.

The saying attributed to Eric Tabarly: 'What you don't have on board won't break,' is certainly true. But on the other hand, if you need something, you really need it. And you really need runners, unless you have a rig with aft raked spreaders.

[1] The 'swept back rig' with aft raked spreaders has the upper shrouds positioned well astern of the mast in such a way that they can do the job of the backstay and/or runners, thus making runners unnecessary and greatly reducing the structural function of the backstay.

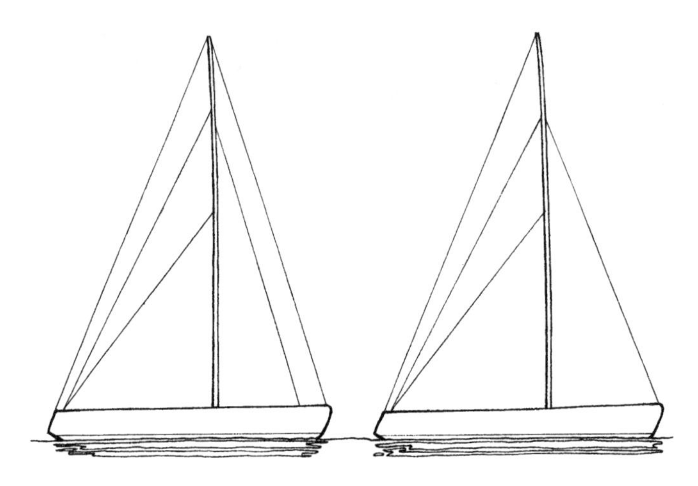

1.1
The difference between a masthead rig (left) where the forestay is attached to the masthead and (right) a fractional rig where the forestay is attached lower down.

On a masthead rigged boat, when you tension the backstay to take the sag – the curvature caused by the pressure of the wind on the genoa – out of the forestay, it is easy to see that a part of this tension (the horizontal component) will bring the masthead aft and thus reduce sag. But unfortunately the tension also has a vertical component that has the negative effect of compressing the mast. And if this compression increases beyond a certain point, it will induce sag even worse than that we are trying to eliminate.

In a masthead rig, the runners[2] allow us to counter the bending of the mast caused by the tensioning of the backstay, while in a fractional rig[3] the lower runners counter the same effect caused by the tensioning of the upper runners. There is another thing that helps explain why runners are necessary. If a masthead rigged boat is sailing hard on the wind with full main and the heavy genoa, and there is a sea running, the mast will be seen to 'pump' (to bend back and forth) with every wave. To avoid this, forward the babystay[4] is tensioned and aft the

[2] We speak of a single set of runners when there is only one line attached to the mast; two sets when there is an upper and lower runner. The upper runner is attached to a higher point on the mast, and also has a higher breaking strain. When there are three sets, there are one upper and two lower runners.
[3] A fractional rig is defined as 9/10 or 7/8 if the forestay is attached close to the masthead or 3/4 if it is attached decidedly lower.
[4] The babystay is a stay anchored to the deck just forward of the mast (on a sloop of 12 m overall the chainplate of the babystay will be about 2 m from the mast) and fixed to the mast about one-third of the way up. It obviously serves as a reinforcement for the mast forward, and also controls and minimises the pumping of the mast when beating upwind with sea running. Moving forward, on boats of more than 14–15 m overall, we have the inner stay and finally the forestay.

runner (usually the lower one). Together, they will hold the mast still and stop this pumping movement.

Things are even worse if on the masthead rig the inner forestay is tensioned to set a staysail. This extra foresail will increase the compression on the mast, and

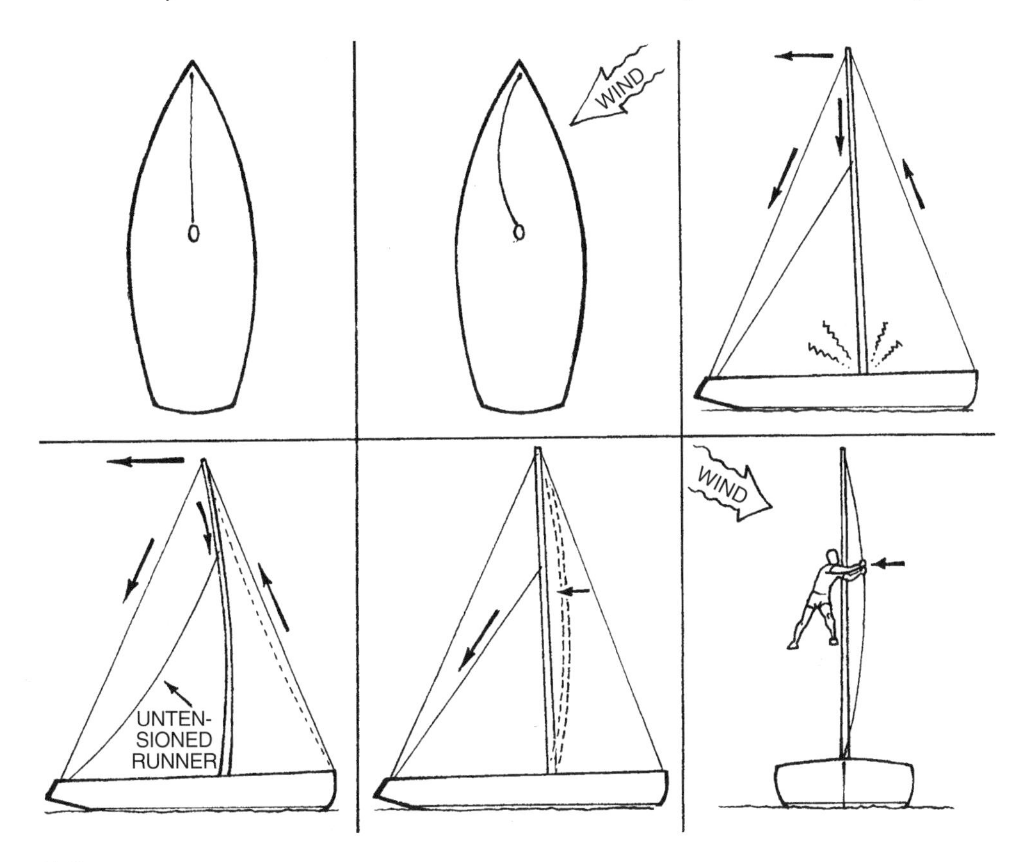

1.2

1. Static situation: at the quayside with true wind speed zero; the forestay is straight and there is no sag.
2. Dynamic situation: under way, the apparent wind speed is 25 knots, the forestay is curved and sag is very pronounced.
3. If we tension the backstay (or the runners on a fractional rig) we tension the forestay, but part of the force has a negative role and compresses the mast.
4. The compression on the mast makes it bend, partly frustrating the positive action of tensioning the backstay.
5. If we tension the runner the mast straightens and thus the sag is also reduced.
6. The ideal would be to have 'someone hovering in mid air and pulling the forestay upwind'.

hence its tendency to pump with sea running, and will require an upper runner leading aft to counteract this. Incidentally, if we were in the land of dreams, the best way of removing the sag from the forestay would be to exert a force equal and opposite to the sagging force at the point of maximum curvature on the stay. How? Well, it is certainly not easy! You would need somebody to hover in mid air while the boat was under way and 'pull' the forestay to windward with the same force, but in the opposite direction, as that with which the sail was making it sag to leeward. In the real world, we prefer to use a line that runs from where the forestay is attached and leads aft where it is tensioned as needed.

On a fractional rig the need for a runner is even greater. The upper runner has the same function as the backstay in a masthead rig: to hold the mast up! In fractional rigs, to underline the vital importance of this piece of rigging, the runner is said to be 'structural'. On rigs of this kind it is not completely unusual to find three sets of runners.

Why open this book by talking about runners? The answer is simple: runners, the kings of running rigging, best exemplify the basic concept of rigging itself, for they are called upon to offer at one and the same time qualities that appear to contradict each other: reliability, speed (both in tensioning and easing) and precise regulation. These are all fundamental characteristics that we will find to varying degrees in every part of a rig.

Runners at Work

It is wrongly said that runners are a hallmark of modern boats, while in fact the opposite is true. Here is a quote from Carlo Sciarrelli's book *Lo Yacht*:

> Olin Stephen's *Circe*, who made her appearance in England for the 1951 Fastnet, had a mast that stayed up without runners, had only one set of spreaders and a single masthead forestay.

So there is nothing new! Again, about 30 years later in the summer of 1983, the English (and not only the English) greeted *Pinta*, a big fractional rig 43 footer owned by German Willi Illbruck and designed by Judel and Vrolijk, with derisive chuckles mixed with disbelief as she sailed into the waters of Cowes. This reaction melted like snow in the sun after the crushing victory of the German team in that year's Admiral's Cup, which dispelled any doubts about the validity of the fractional rig for racing yachts.

The fleet of IOR 30.5 raters in the 1984 One Ton Cup, held in Trinité-sur-Mer, France, was made up almost entirely of fractional rigs, and marked the definitive affirmation of this type of rig on the racing circuit, even on big boats.[5] So we can use the One Tonner as a testing ground for the observations we will be making.[6]

The upper runner of a fractional rig of this kind has a working load of about 2200 kg. Note that this value was obtained through field measurement, with a tension gauge with a load cell mounted on the stay: in 1984 on the One Tonner *Brava*, a beautiful Vallicelli built like a Stradivarius by the Morri & Para yard, we had a hydraulic tension gauge! The stay on these boats had a maximum working load of 3400 kg.

We will explain later how to determine the working load on the various pieces of rigging, using special formulas and load deviation angles. Today, load cells are all electronic. Here it is vital to underline the importance of our starting point. It does not matter whether this comes from practical experience on board or from theoretical calculation. What does matter is that we must set out having a very clear idea of what loads our piece of rigging must bear. For example, too many people still think that the load on the number 1 genoa sheet is greater than that on the sheet of the number 3. In fact the opposite is true (we will look at this in more detail later).

It is vital that the starting point be clear, from both a pragmatic and a logical standpoint (though we are well aware that the science of boating knows above all that it does not know everything!) Otherwise we would set out, right from the very start, on a road that would lead us to wrong conclusions. To have a sufficient safety margin with respect to our working load of 2200 kg we should use, for the runner, a line with a breaking strength not less than 3500 kg.

[5] It is well known that the fractional rig was favoured by the New Zealand school led by Farr (Davidson and White, among others), but an internationally famous yacht designer, with several America's Cups to his credit, predicted well before the event the great flight of owners from racing yachts, which led to the incredible cancellation of an edition of the Admiral's Cup, caused by the extremely radical changes to the class which was very uncomfortable to sail because of its fractional rig. It was no accident that the *coup de grace* to the Admiral's Cup came after the organisers decided the Fastnet fleet should be composed of the small, fractional rigged Mumm 36. And it is no accident that today the big revival of owners in classes like the Farr 40 is due in part to the fact that these boats have no runners. An identical thing happened in windsurfing with the advent of the *funboard* that brought radical and extreme changes to the category and undeniably frightened away a huge number of devotees.

[6] To anticipate any objections, I would like to point out that speaking today of an IOR One Tonner of 20 years ago is neither obsolete nor *démodé* . . . the loads present on those boats are identical to those on today's IMS 40 footers and – strange but true – on certain very interesting 44 foot day-cruisers, designed, like dinghies, for outings of just a few hours! They are super light and have a particularly low righting moment.

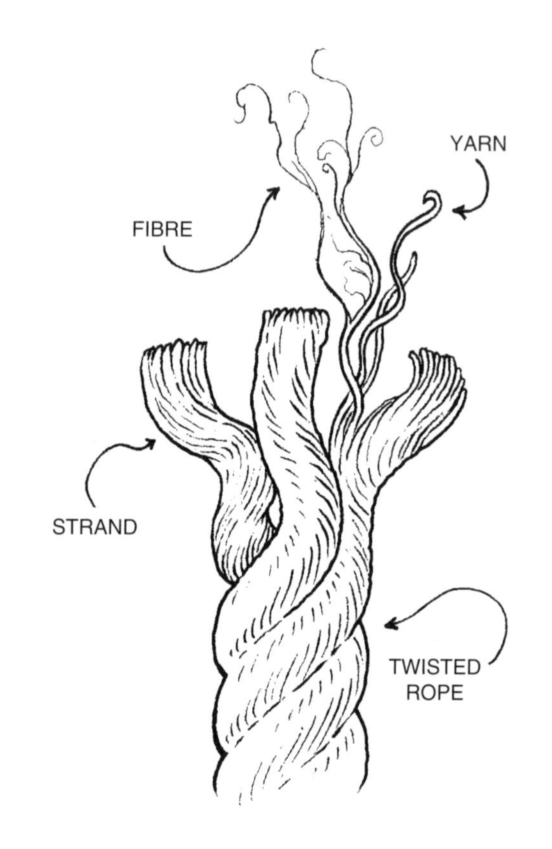

1.3

An example of three-strand rope. First stage: the fibres are twisted to form yarn. Second stage: the yarn is twisted into strands. Third stage: the strands are twisted. Fourth stage: the twisted rope is made.

We say 3500 kg to have a minimum safety coefficient[7] of 1.5, which allows us to have a breaking strength that is 50% more than the maximum working load we expect on the runner. A choice that allows us to sleep at night. On a cruising yacht, we would certainly prefer to use a safety coefficient of 2, which would give us a breaking strength 100% greater than the presumed maximum working load, though the weight of the line would be far greater than that chosen for the racing yacht.

For love of paradox, but above all to help understand why we choose one material rather than another, let us imagine for a moment rigging our One Tonner exclusively with materials that were widespread in the first half of the past century. We would have used laid three-strand tarred hemp rope (similar to today's mooring lines) for the runner. But using this would mean, at the effective working

[7] In our case it is 1.6.

load of 2200 kg,[9] that the rope would stretch several metres. And we could not tolerate this, for once the effective working load was reached the runner would stretch so much – like an elastic band – that we would have to tension it further, and this would stretch it even more, so we would have to tension it again, thus stretching it even more again. And this would lead us into an unending vicious circle and we would lose our patience and perhaps also our mast!

There is no doubt that the choice of hemp rope satisfied our first fundamental requirement in terms of breaking strain. But that is certainly not enough. For if the breaking strain were assured, and the legitimate need for light weight and low cost were met, we would still end up with a miserable failure. The rope would be so elastic as to be completely unusable, and we would be at the danger limit.

Let us learn here that rigging is the art of compromise: you gain on one side and lose on the other. The perfect balance of the various components of a piece of rigging, meeting the requirements in this way, determines the success of a project, and thus of the project put into practice.

But let us continue with our hypotheses. Since '. . . the first metallic shrouds, in galvanised wire, appeared on the cutter *Cymba*, built by William Fife in 1852 . . .',[8] at this point let us try using a 19 strand metal wire in 316 stainless steel, external diameter 7 mm and breaking strain 3550 kg.[10] To check whether our current choice is an improvement as concerns the problem of excessive stretching we encountered with hemp rope, let us right away calculate the stretch. The runners are 15 m long and each weighs 3.645 kg. The elastic stretch in millimetres is given by the formula $W \times L/E \times A$, where W is the load applied in kN; L the length of the wire in millimetres; A the area of the cross-section of the wire derived from the formula $D^2 \times 3.14/4$ and E the module of elasticity for the material based on its specific composition. So we will have: $21.5 \times 15000/107.5 \times 38.46$, which gives us a stretch of 78 mm.[11]

We have certainly made a lot of progress in reducing stretch compared with the three-strand tarred hemp rope: from metres of stretch we are down to

[8] C. Sciarrelli, *Lo yacht*, Mursia, p. 49.

[9] The weight does not include the terminals.

[10] The breaking strains vary from maker to maker. Here for the 19 strand spiral wire and the rod we refer to the 1989 Riggarna catalogue.

[11] Total stretch is the sum of permanent and elastic stretch. Permanent stretch comes about simply because the fibres and/or strands of which the line is made become more compact and uniform along the longitudinal axis of the line when the first loads are applied. This stretch is permanent, and will be there for the entire life of the line. Elastic stretch is typical of all metals and tends to return to zero once the load is removed. Its value increases with load, according to Hooke's law. Resistance to stretch is known as the elastic module of the material: rather than using the classic Young's module – to compare different kinds of line – here we talk about the elastic module of the line according to how it is made up.

Table 1.1 Typical values for *E* in kN/mm² for various materials

spiral 1 × 19	107.5
dyform 1 × 19	133.7
nitronic 50	193.0
Kevlar rod	124.0
wire 7 × 19	47.5

millimetres! But when we get to our fateful One Ton Cup[12] we realise that many of our most fearsome opponents are already using for their runners another material that is decidedly more advanced: a single rod of steel known as Nitronic 50. The diameter of rod that comes closest to our needs is 5.5 mm with a breaking strain of 3220 kg.[13] The formula we used earlier gives a stretch of 70.3 mm. So we have slightly lower stretch than the spiral wire, though not significantly so, just a few millimetres. What is significant, however, is the weight saving: from 3.78 kg with the spiral wire we are down to 2.85 kg with the rod.

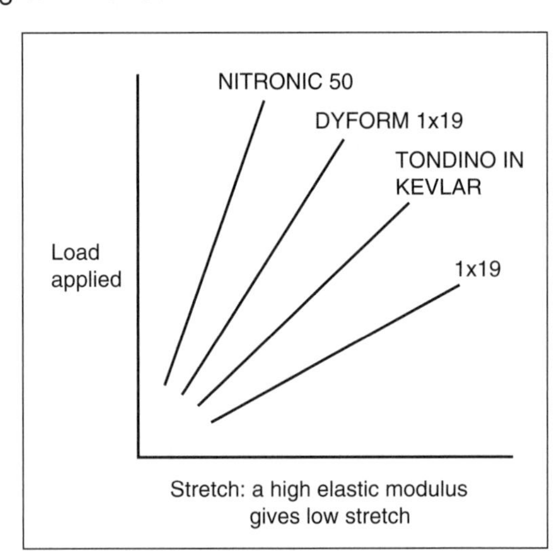

1.4
A diagram showing the relationship between the elastic modulus of various materials and kinds of wire and the resulting stretch.

[12] The One Ton Cup is the world championship of the One Tonner class. The regatta has been held for very many years.
[13] Note that there is a slight reduction of the security coefficient, which with rod becomes 1.45. Usually on racing boats it is 1.5. In fact it would be preferable to use Newtons – abbreviated to N – instead of kg as units of measurement for kilogramme forces. If you do not have a calculator you can roughly convert N to kg. by dividing by 10, for example 10 000 N roughly equal 1000 kg. With a calculator, divide the N by 9.81 to get kg. Sometimes you find dN (decaNewton) which are, with dockside approximation, equal to kg., though to be precise you divide them by 0.981 to get kg. Values in kN should be divided by 0.009 81 to get kg.

We must bear in mind, too, that this saving must be multiplied by two, as runners are always in pairs, and above all note that the weight we have saved would have been at a height of about 7.5 m above the deck, with all the negative effects it would have had on heeling, pitching and rolling.

But we must also underline a disadvantage for each of the last two options. And this will allow us to introduce a third and very important factor that we really must take into account in choosing a material and the form of that material for a piece of rigging. Besides breaking strain, working load and stretch, there is practicality.

Unfortunately spiral wire has an annoying tendency to twist under tension. And it is no small problem when you realise that the block of the runner you are tensioning has rotated and taken a couple of twists. This makes it hard to tension the runner further or even to ease it. This is a nasty problem and unfortunately we have to take it into consideration when we have a runner in spiral wire that passes through a block. If the wire ends without a block, and thus the runner goes directly to its tensioning system, a winch or similar, the defect is still present but is obviously less damaging. Rods, since they are formed of a single piece, are 'stable' and so do not have this tendency to twist. But they have another disadvantage that is no less serious: they are particularly susceptible to knocks from the boom when gybing. These knocks bend the rods, thus shortening their working life and reducing their strength.[14]

Another hypothesis that tends to reduce both these disadvantages is to use AraLine 49, a braided Kevlar[15] line with a protective polyurethane outer sleeve. These lines were distributed in the boating sector by the French company EPI, now marketed by Navtec.[16] AraLine 49 with an external diameter of 10 mm has a breaking strain of 5184 kg, and at the working load of 2200 kg stretches by 0.49%. Over a length of 15 000 mm, that means only 73 mm of stretch.

Certainly this is more stretch than in the case of steel rod, but we have a decisive saving of weight (each runner would weigh 1.335 kg) and above all we eliminate the practical problems we had with both spiral wire and steel rod, for *Araline* does not twist under tension and is well able to cope with any knocks from the boom.

Let us sum things up at this point. Though it may seem paradoxical, the first solution, with three-strand rope for the runners, is the typical one for very low

[14] You may cast the first stone only if you are sure that you have never, during a gybe in a blow, sent the boom crashing into what had become the lee runner: the one that is tricky to ease and is thus usually eased late!

[15] In most cases Kevlar 49 is used, while for more sophisticated applications such as the Volvo Race Kevlar 149 has been used; this has an elastic module 30% greater than that of the 49.

[16] Today runners of this kind are made not just by Navtec but also by the American company Aramid Rigging, the English makers Rig Shop and Future Fibres and Italy's Gottifredi & Maffioli.

budget sailors. The Bohemians of the seas, the hippies of the ocean certainly do not disdain it, and if you find yourself in one of the ports they frequent you will see several examples of it. Runners in 19 strand spiral wire are usually found on former racing boats adapted in a makeshift way for sporty cruising, while runners in exotic fibres are found exclusively on racing boats and very high prestige cruising yachts. All have clear and distinct pros and cons, as we have seen, but the only solution I really would recommend abolishing is the steel rod, for its disadvantages far outweigh its few advantages.

What I am saying is that the ideal solution in absolute terms does not exist; what does exist and has a meaning is the '*xy*' version that is the best solution available for a given problem.

Table 1.2 Performance comparison between various kinds of runners

Material	Reliability	Elongation	Handyness	Life span	Weight	Price
3 strand hemp line	low	very very high	low	medium	low	very very low
1 × 19 316 s.s.	very very high	medium	very low	medium	high	low
Nitronic 50 rod	medium to low	low	very low	medium to low	medium	high
Unidirectional Kevlar fibers EPI/Navtec cable	Medium to high	low	very high	low	low	high
As above but made in Vectran	high	low	very high	medium	low	high
As above but made in PBO	high	low	very high	medium	very low	very high
Continuous loop cable (Kevlar, Vectran or PBO made) built by Aramid Rigging or Chien Noir, with thimble-like terminals	high	very very low	very very high	high	very very low	very very high

Table 1.2 *Continued*

Material	Reliability	Elongation	Handyness	Life span	Weight	Price
12 yarns spectra single braid made by Samson (Spectron 12 plus)	very very high	medium	very very high	very very high	very very low	medium
As above but with Vectran (New England v12 or Yale Vectrus) or aramidic (Samson Technora)	high	medium	very very high	high	low	medium

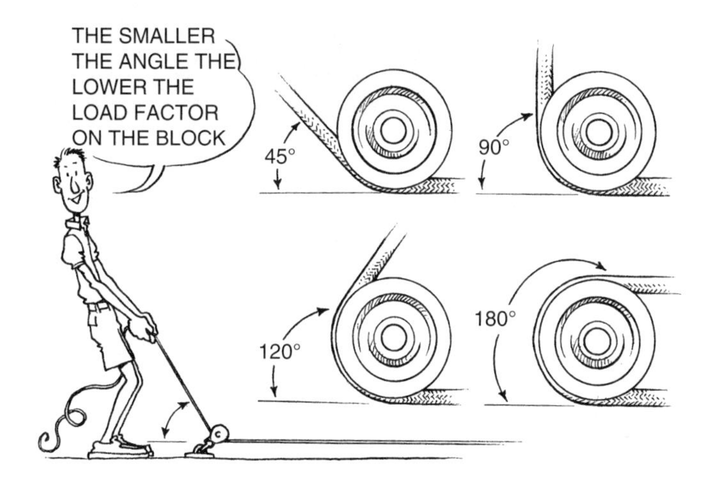

From the EPI style runners of the 1980s to the present day we have seen a series of attempts to improve things that have not always succeeded perfectly from all points of view. First of all, work has been done on the quality of the material used for the terminals of the runners, moving from normal 316 steel to 6061 aluminium and, where class rules permitted it, titanium. But the most extensive experimentation has been carried out on the material used for the core and its morphology. With the advent of fibres of the latest generation, Kevlar has been replaced both by Vectran and by the very costly PBO to obtain better mechanical resistance and tolerance of solar radiation and bending, but with the disadvantage of astronomical expense.

1.5
Yale's Vectrus single core line, used for a spinnaker guy.

But the biggest breakthrough has come from the manufacture of these lines no longer with the classical technique of parallel fibres or fibres braided like those used in double braided line, but instead – once the overall pin-to-pin length[17] of the line has been established – by winding a very tiny thread of the chosen exotic material around this length until the breaking strain required by the project has been achieved. It is easy to understand that once this winding has been done, the knot or braid used to make fast the two free ends of the thread has to support only a very tiny part of the working load, so this manufacturing method is much more reliable than the classic splicing.

Also, this unending thread does not need heavy terminals like those of one-way lines but only thimbles, though sometimes these must be of a special kind.

[17] The 'pin-to-pin' length is the length of a piece of standing or running rigging between one anchor pin and the other. For example, in the case of the forestay, between the pin of the bow chainplate and the masthead pin.

1.6

Left: a classic soft eye splice, that is without a thimble, on a double braid line, used in upper and lower runner systems with an eye more to cost than to performance.

Right: an upper runner in exotic fibre and lower runner in wire rope, connected by a swivel plate.

Runner Blocks and Their Circuits

When and how to avoid them and, if you really cannot do without them, how to size them

Let us be honest, even without heeding Eric Tabarly's warning, we really would like to do without the runner block. By runner block we mean the block fitted to the runner cable and usually called the *flying block* because it flies back and forth over the deck, frequently clouting the helmsman over the head on its way, depending on whether the runner is in tension or not. A *deck block*, on the other hand, is fixed to the deck. This block is generally bulky, heavy, costly and in some cases downright dangerous.

The maxi *My Song*, a Reichel-Pugh 85-footer, when racing keeps her runners close to the centre of the transom, but when cruising they are moved to points out on the quarters so as to avoid the flying block getting in the helmsman's way when the runner is eased.[18]

[18] In truth, runners outside the stem-stern axis of the boat, once tensioned, also apply a lateral twist component to the mast, besides that acting aft – to windward, and this tends to work against the upright position of the mast, so when possible it is preferable to have runners anchored as close amidships as possible.

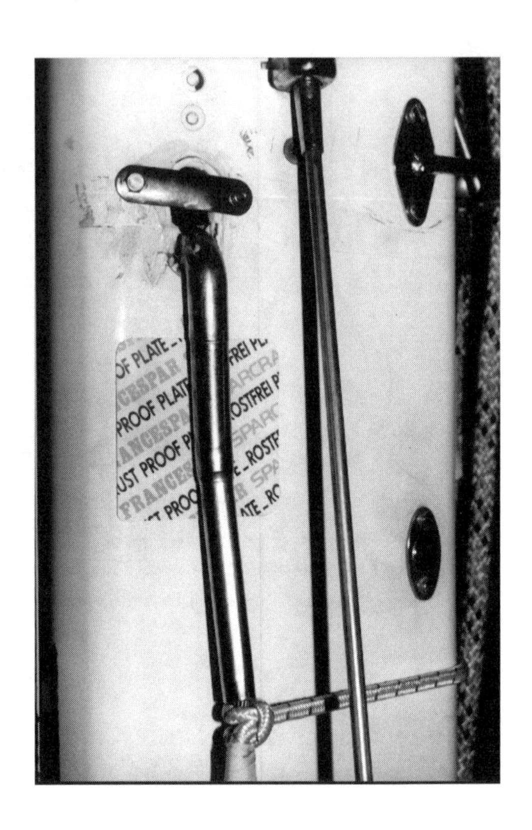

1.7

The 'T' terminal of a 1×19 strand spiral wire runner. Note the safety plate for the 'T' terminal and the shock cord.

These observations help us understand what we want from the flying block: we want its dimensions to be as little gigantic as possible, and it should be as light, inexpensive and reliable as it can be. Having said that, we quickly realise that we are yet again in that age-old and intricate situation of finding the best compromise from among the various solutions to the needs we have described. We will see later how to unravel the problem, but meanwhile let us examine an apparent shortcut and take a look at a runner circuit that does not use a block on the runner itself.[19] The only runner circuit that does not need a block fitted directly to the runner is the 1 : 1 circuit where the runner (either a single runner or an upper and a lower runner) leads either directly or via a link to a winch or other tensioning system. Let us examine the pros and cons of a circuit of this kind.

[19] The runner at rest – that is when it is not needed, in port, or simply because it is the lee runner and thus has to be eased – lies close to the shroud chainplates or – only in the case of a masthead rig where there is a backstay where a small block can be fitted – along the backstay, held close to it by a powerful shock cord rove through this small block.

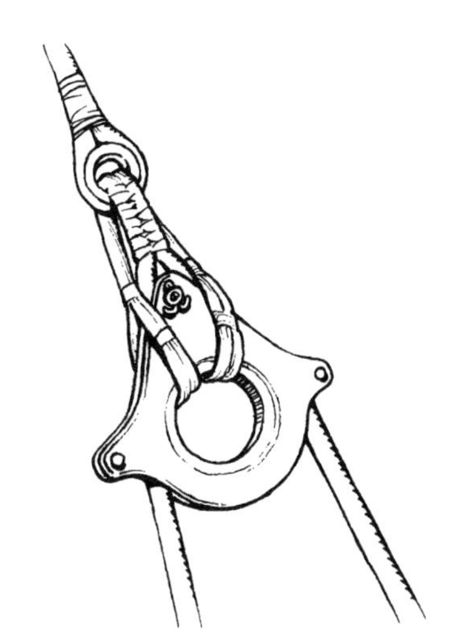

1.8
The flying block of a 2 : 1 runner. Note the special sailmaker's thimble at the extremity of the upper runner in continuous thread; the lower runner is fixed to the block with a strop made of several turns of Vectran between the sheave of the block and the runner.

The pros are easy: if to tension the runner you have to haul, for example, 5 m of line because that is the distance from the shrouds, where the runner usually rests, to the block that turns the line, if there were a 2 : 1 system you would have to haul double that amount of line, that is 10 m. The 1 : 1 system is far faster both in tensioning and in easing. The 2 : 1 system with its flying block offers a lot of resistance to the necessary retaining shock cord of the runner that needs to be eased, both because of the weight of the block itself and because of the friction of the line that has to run for a full 10 m through the blocks in the system.

In short, a direct circuit with no blocks, as is the 1 : 1, has as its (only) strong points simplicity, speed and practicality. On the other hand, its downside cannot be ignored. When you haul the runner directly, with no block and tackle interposed, the load on the block on the deck is extremely high, so you need a very robust block (and that also means you need to have a sturdy reinforcement on the part of the deck where the block is fitted) and you will also need a very powerful winch or other tensioning system. And let us make it quite clear that 'robust and powerful', while sounding reassuring to many sailors, also means high weight and cost, and these are negatives. In practical terms, let us sum up the pros and cons of this kind of circuit:

- *Pros* – very practical to use both for tensioning and easing the runner.
- *Cons* – needs a very robust block; demands a sturdy reinforcement on the part of the deck where the block is fitted; needs a winch – or other tensioning system – that is particularly powerful.

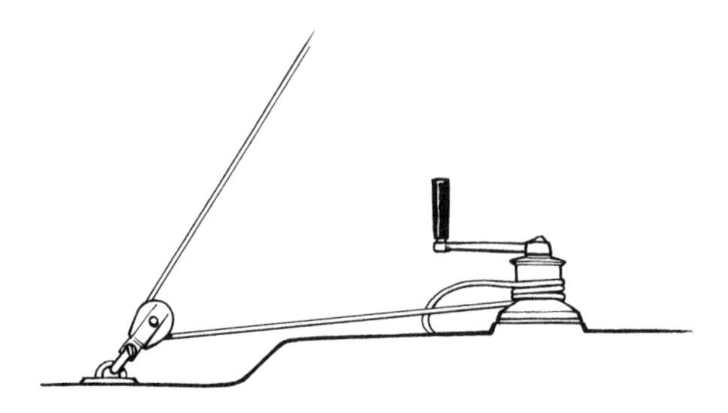

1.9
Example of a 1 : 1 runner circuit.

Let us look closer at the issue with the help of the drawing above. The runner is in 19 strand spiral steel wire 5 mm in diameter with a tail in fabric line that ends up on the winch; sometimes a stopper is placed between winch and block to free up the winch for other uses. But what we particularly want to underline here is the calculation of the breaking load of the block in a circuit of this kind. We might at first think that the block on the deck should have a breaking load equal to or greater than that of the wire rope used for the runner. The wire rope in question has a breaking strain of 2130 kg and it could occur to us at first sight to choose a block with the same breaking strain. Nothing could be more mistaken! In fact, if for any reason the breaking strain of the wire (2130 kg) were reached, the block (with the same breaking load) would have given way much early, for the angle (of 110°) at which the block turns the line to the winch causes an increase in the load of 64%!

So the calculation we need to do is this: breaking load of block = breaking load of line (or wire rope) multiplied by the load increase factor (a factor that depends on the angle by which the line is turned). That is: breaking load = 2130 × 1.64 = 3493 kg. In this way we have introduced one of the cardinal principles of dimensioning rigging: the increase in load due to the angle by which a line has to be turned.

Table 1.3 Table showing the more common turning angles and the relative load increase – or decrease! – factors

Deflection angle	Factor	Deflection angle	Factor	Deflection angle	Factor
30°	52%	90°	141%	150°	193%
45°	76%	105°	159%	160°	197%
60°	100%	120°	173%	180°	200%
75°	122%	135°	185%		

There are three interesting things to take special note of in the table above. First of all, we must pay special attention to all the blocks (and sheaves) that turn lines through 180°: we are talking about the classic blocks placed astern for the spinnaker sheets or the masthead spinnaker halyard blocks (on those masts that have attachments for an external spinnaker halyard). These blocks must support twice the load on the line. So take care to verify their breaking (and working) load and size them accordingly!

All blocks or sheaves that turn line by an angle of 60° bear the same load as that on the line, no more and no less, and should be sized accordingly. All blocks and sheaves that turn the line by less than 60° bear a load less than that on the line, and should also be sized accordingly.

Let us now directly compare a 1 : 1 circuit for the runner tail with a 2 : 1 circuit. Let us make it clear once and for all that the load reduction capacity of a tackle is determined by the number of parts of rope entering or leaving the moving block.

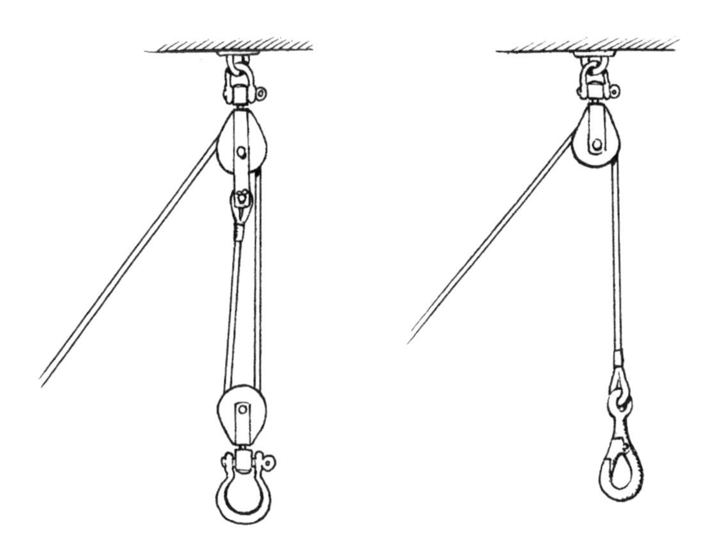

1.10

Left: This drawing may seem a 3 : 1 purchase but is in fact a 2 : 1, since only two segments of line leave the lower block (the actual mobile part!) and thus the load is halved.

Right: In the drawing alongside, although the line is rove through a block the resulting purchase is only 1 : 1 and so there is no load reduction. It is simply easier to use.

Besides having a suitable breaking load, the flying block must have a series of characteristics that set it apart from other blocks, which only in appearance seem suited to the job. Besides being light and tough, it must have lateral guides for the sheave so that the runner tail, when eased, does not jump off the sheave, for this would be dangerous: when the runner is tensioned again it may cut through the tail and/or the tail may end up between the sheave and the cheek, thus seizing up the block. It should also permit the fitting of a special attachment so that it can be easily hooked up both to the upper and lower runner. A suitable block would be one having these characteristics and with a breaking load of 4400 kg and a working load of 2200 kg.

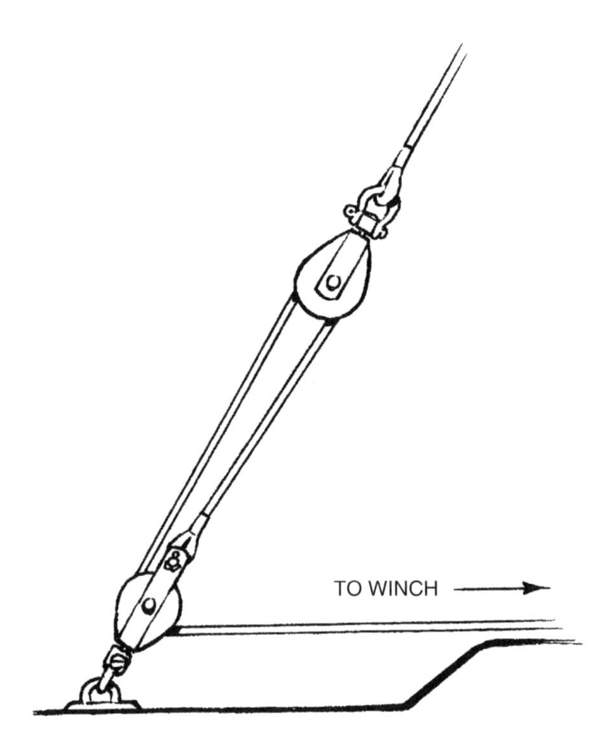

1.11

Say the wire used for the upper runner (and for the lower runner) has a total breaking load of 4400 kg. The upper block we fit to the runner must not have a breaking load lower than this, for the simple reason that in a chain no link must be weaker than another, otherwise the chain will break at its weakest link.

Note one thing. With upper (and lower) runners in exotic fibre you must bear in mind that if you use them to replace those in wire rope or rod (metal) and want to have the same stretch, you will find they have a breaking strain much greater than that of the metallic runners, so the rule described above is no longer valid – it would lead us to having enormous blocks. The solution is to compare the exotic fibre runner – with the same stretch – with its metal counterpart and choose a flying block with the same breaking load as the wire rope (or rod) shown in this comparison.

Table 1.4 Specifications of exotic line for runners in aramid fibre Comparison with 1 × 19 wire rope in SS316 with the same stretch

Exotic Line							1 × 19 WIRE					
Line code	External sleeve diameter (mm) (in)		Minimum breaking strain (kg) (lb)		Weight (kg/100 m) (lb/100 ft)		Wire diameter (mm) (in)		Minimum breaking strain (kg) (lb)		Weight (kg/100 m) (lb/100 ft)	
0.8T	4.5	0.18	800	1760	1.60	1.10	3	1/8	720	1584	4.49	2.95
1.5T	6.0	0.24	1500	3300	2.80	1.90	4	5/32	1280	2816	7.81	5.25
3.0T	8.0	0.31	3000	6600	4.90	3.30	6		2880	6336	17.60	11.80
3.8T	8.5	0.33	3800	8360	5.50	3.70		1/4	3220	7084	19.40	13.00
5T	9.9	0.39	5000	11000	7.40	5.00	7	9/32	3550	7810	23.90	16.10
7T	12.0	0.47	7000	15400	10.90	7.30	8	5/16	4640	10208	31.20	20.90
9T	13.1	0.52	9000	19800	12.80	8.60	10		7250	15950	48.80	32.80
12T	12.0	0.59	12000	26400	16.80	11.30	12		10400	22880	70.30	47.20
15T	16.8	0.66	15000	33000	21.10	14.20		1/2	11650	25630	79.30	53.30
20T	19.1	0.75	20000	44000	27.10	18.20	14	9/16	14180	31196	95.70	64.30
25T	22.0	0.87	25000	55000	35.90	24.10	16	5/8	18560	40832	125.00	84.00
31T	25.0	0.98	31000	68200	46.20	31.00	19	3/4	21260	46772	176.00	118.00
43T	28.9	1.14	43000	94600	51.30	41.20	22	7/8	29070	63954	236.00	158.00
54T	32.3	1.27	54000	11880	76.30	51.30	26	1	40600	89320	330.00	220.00

Meanwhile the lower block – the one on the deck – should be of what kind? And with what breaking load? Here we need that masterly tool that is the table of load factors in relation to angles. We have seen that a block that turns a line through an angle of 180° has to support a load equal to double that on the single line. Applying this logic in reverse, in the case of the drawing of the 2:1 tackle, and bearing in mind that the elements known to us are the breaking strain of the line (4400 kg) and the angle of deviation (180°), the maximum breaking strain we can have in each length of the runner tail is 2200 kg, because the block halves the load on each single length.

But that is not all. Because while one of the two lines coming from the flying block ends up on the head of the deck block, the other line is turned by this same block towards the tensioning system, in this case a winch. As we have seen, a deviation of more than 60° increases the load on the block with respect to the load on the line, while one of less than 60° reduces it. The deviation in question is 105°, which gives a load factor of 1.59. Thus we have 2245 kg × 1.59, giving a load of 3498 kg, and to this we must add the other 2200 kg of the line that is rove to the head, and the final result is a minimum breaking load for the block of 5698 kg.

Thus we have defined the form and loads of the deck block: we need a block with a swivel, a single sheave and a becket[20] with a minimum breaking load of 5698 kg. The block will need a pad eye to fix it to the deck, and that makes it tend to lie down when the tail of the runner is eased, unless we keep it upright – stand-up, as they say[21] – with a piece of shock cord that connects the block to the rail.

The deck block could reasonably have a 100 mm diameter sheave and weigh 750 g (without counting the pad eye that will be needed).[22] Thus its weight, dimensions and breaking load are greater than those of the flying block, and this is another reminder of how careful we must be about the angle by which a block turns a line, so that we can dimension blocks correctly.

Now we have completed the dimensioning and the resulting selection of the blocks for this 2:1 runner tail circuit, so-called because it halves the load of the runner that ends up on the winch (or tensioning system). But we can rightly ask: when do I choose a 1:1 circuit and when should I opt for a 2:1 circuit?

To answer the question, let us take a step back and return to the boat we are using as a testbed: the runner of the *One Tonner* had a working load of 2200 kg.

[20] We will speak of a becket when a purchase begins or ends at a fixed point: this may be a simple eye on the deck or an attachment point for a block.

[21] A 'stand-up block' is one designed to remain permanently upright.

[22] Actually this is not strictly accurate! A block, even if it has bearings, causes friction, and this increases the load on a line passing through it.

Now, if we want to tension a runner with a working load of 2220 kg, using a direct circuit with no reduction, that is a 1 : 1, we will have to use a winch with a power ratio – at the lowest speed – of 74 : 1, and taking into account an efficiency for the winch of 75%, to get the result we want we will have to exert a force of almost 40 kg on the winch handle with every turn. And that is no small effort, even for a fit person.[23]

A winch of this kind could be the Harken[24] 74.2 STR, which weighs 27.2 kg. But it has to be said right away that not only is this not the right winch for the job, but also that it is totally a mistake to try using a 1 : 1 circuit for the runner of a One Tonner.

In other words: a direct circuit demands – because of the high loads involved – a very powerful, and thus very expensive and very heavy winch. In fact you get the best compromise between power and speed in a runner tail circuit that has to handle a working load of about 2000 kg by halving the load of the runner with a block between it and the winch, which in our example leads us to use a winch with a ratio of 48 : 1 (with a load of 40 kg on the handle), and this corresponds to a Harken 53.2 STR winch weighing 12.5 kg.

Table 1.5 Table comparing 1 : 12 systems with 2 : 1 systems, especially as regards weight

	1 : 1 system	Weight (kg)	2 : 1 system	Weight (kg)
Winch	2 Harken 74.2 STR	54.4	2 Harken 48.2 STR	21.0
Flying block	–	–	2 Harken h 1991 – 75 mm	0.48
Deck block	2 Harken h 3007 100 mm	1.14	2 Harken h 3007 100 mm	1.28
Total		**55.54**		**22.76**

The weight saving on two winches with this latest solution is 19.4 kg, compared with the 1 : 1 direct circuit or 28.78 kg, if we take into account, as we should, that in the 2 : 1 circuit there are additional blocks. It is clear that the 2 : 1 system is far superior to the 1 : 1. Not least because the 1 : 1 would cost almost three times as much and weigh more than double!

And as if this quality/price advantage were not enough, we would also like to point out that the 2 : 1 system has on its side greater practicality, because when you need to ease the runner tail – since it is under only half the load of the runner itself – it is much easier, whereas with the other system the load on the runner tail is high.

[23] Power ratios and other matters relating to winches will be dealt with in a later chapter.
[24] Winches with this power ratio are also made by *Lewmar*, *Antal* and others.

And this high load means that when you ease the runner, the tail jerks out, making the mast oscillate back and forth like a whip, and you also risk 'searing' – as if you had passed a hot iron over the line sleeve – the runner tail. However, today there are lines with special sleeves to prevent this searing, but apart from the matter of very high cost, the problem of the whiplash effect remains.

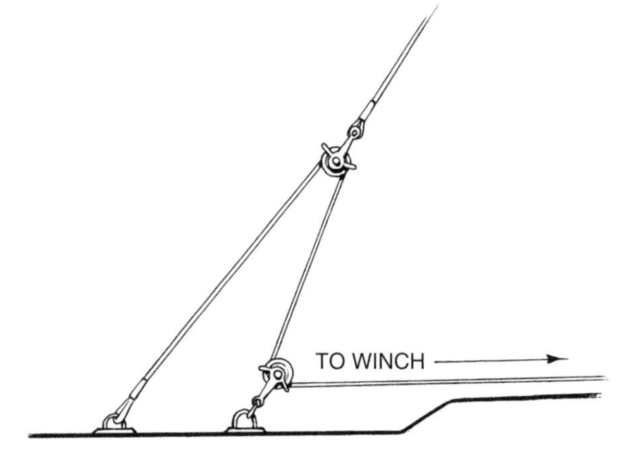

1.12

An improved version of the 2 : 1 system with a runner circuit with a 2 : 1 purchase and the tail fixed to an eye separated from the deck block.

The system illustrated on this page is again a 2 : 1, but here the tail is not rove through the deck block but fixed to an eye on the deck itself. This has basically two very important advantages with respect to the 2 : 1 we saw earlier with the line rove through the deck block. Since the deck block here does not bear the load from one part of the runner tail (which is no longer rove through it!) it can be smaller. The distance between the deck block and the point (eye or pin) to which the tail is fixed forms a 'base', and the longer this base is the more it guarantees that the runner tail will not twist when it is tensioned.

Having said this, we will now go into greater detail on the fundamental concept that, as the boat we are concerned with gets bigger, an increasing number of blocks will be used for the runner tails. The greater the working load on the runners, the greater the reduction required, for it is indispensable to have a manageable working load on the part of the runner tail that ends up on the winch. And obviously we must choose the right winch to give the runner the tension needed.

The working load of a runner tail is manageable when:

a) you can ease the runner tail, and thus the runner itself, in a hurry without searing the tail because of the high load it is under;

b) you can ease the tail – and hence the runner so as to adjust the tension of the forestay, the mast bend and thus the shape of the sails

– without sudden jerks that would lead to abrupt changes in rig tension.

In the America's Cup, working loads of simply astonishing proportions have been reached on the runners, with 15 000 kg on the upper runner.[25] And in fact the runner tails are handled with a 3 : 1 reduction and a winch giving 100 : 1 reduction, such as the Harken 990.3 STAC.

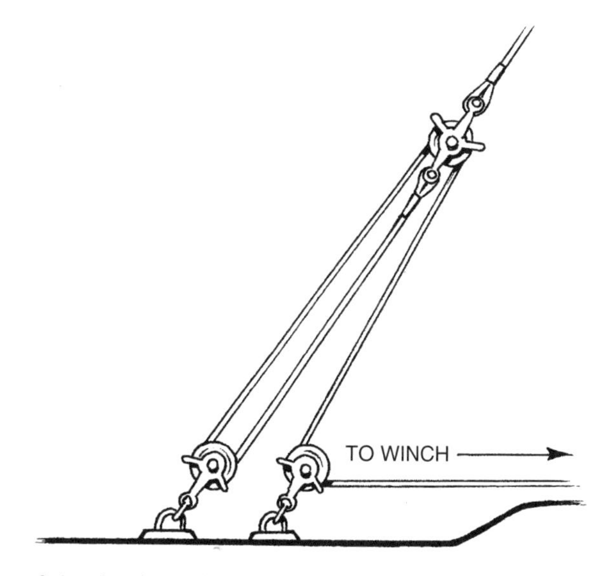

1.13
3 : 1 runner purchase system.

Let us now look at a system of this kind applied to our case and consider together the dimensioning of the blocks (see drawing above). The upper block attached to the runner must – usually – have at least the same breaking load as the runner itself: 4490 kg. This block divides the load into three equal parts and thus the afterdeck block, which turns the tail through 180°, bears twice the load of a single section of the tail, that is 2993 kg, while the forward deck block, which deflects the runner tail through 110° with a consequent load increase of 65%, needs a minimum breaking load of 2469 kg.

But to keep our feet firmly on the ground, it has to be said that while a 3 : 1 allows very high loads on the runners to be managed, we must also bear in mind that such a reduction means we have to haul a full three times the length of line compared with a 1 : 1 system. And this implies that the 3 : 1 system – with very high loads – is destined for use by particularly expert crews.

[25] Actually it would be more proper to speak of the load on the upper runner, the two lower runners and the 'topmast' – a kind of mobile masthead backstay – which had to be double because the cut of the mainsail would not allow the leech to pass below a traditional backstay.

1.14

3 : 1 Frederiksen block system used on the Vor 60 *Illbruck*, previously sailed by Paul Cayard, for the round the world race.

1.15

3 : 1 runner circuit on the J class *Endeavour*.

The Various Systems for Tensioning the Runners

We can have the following tensioning systems for the runners:

- By hand without block and tackle, as on small keelboats.
- By hand for the fast recovery part and then a purchase with aluminium housed sheaves[26] for fine tuning: for small fixed-keel cabin boats.
- By hand but with the help of a tackle for runners on cruising boats and on masthead rigs, where the runners are not structural and need only to be lightly tensioned to counterbalance the inner forestay.
- By hand with the help of levers giving purchase, as on classic boats.
- By hand for the fast recovery part – sometimes also with the help of a purchase – with a winch for fine tuning, as on deep water and America's Cup class boats.

1.16

Harken deck block in the 3 : 1 runner system of a 60 footer (note the safety strop between the chainplate pin and the block itself).

[26] A 'Magic Box' is a self-contained purchase in a square section aluminium housing that has the advantage of giving a lot of purchase without the classic defect of such arrangements, that is the twisting of the line sections and resulting friction, since the tackle operates within guides on the sides of the metal housing.

Let us take a closer look at the lever system, as it is the only one we have not yet examined:

> . . . Chris Ratsey's *Evenlode*, the beautiful fifteen metre double-ender . . . racing in the Solent in 1948 had her runners tensioned by hand using a block and tackle system, because her owner considered levers too dangerous.[27]

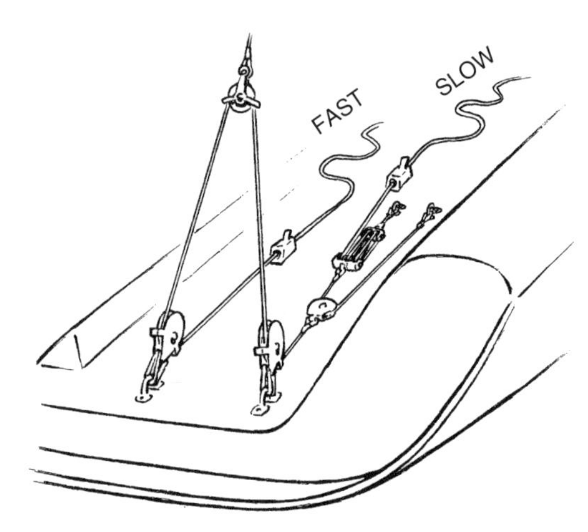

1.17

A 2 : 1 runner purchase for boats up to 30 feet with fast regulation, blocked by a stopper, and fine tuning with a 24 : 1 cascade purchase.

Certainly the owner of *Evenlode* has my every sympathy: when I was mariner on *Sagittario*, a 16 m Sciarrelli cutter that the Italian Navy had fitted out for the Ostar race, and which in my day was at the disposal of the navy head of Sailing Sport for racing and cruising, I just could not bear those blessed levers that I had to carry forward to tension the runners. To tell the truth, levers have in their favour an extreme simplicity of the runner circuit, for the runners generally end directly – 1 : 1 – with a tail usually in flexible steel that passes through a block on the deck before arriving at the tensioning lever. The downside, as mentioned, is that they are clumsy and difficult to handle, though it must be noted that on classic yachts the lever system is often used out of respect for the original rig of the boat.

The Runner Tail[28]

Bearing our example in mind – the runners of a One Tonner – let us try to choose the most suitable runner tail for a 2 : 1 circuit, that is with the following requirements:

[27] Roger Marshall, *Designed to Win*, Granada Publishing Ltd, 1979.

[28] The runner tail is the line that tensions the upper runner (and the lower one, if present).

- *An adequate breaking strain.* Not less than that of the other elements making up the system, for example the flying block;
- *Minimum possible stretch.* As little lengthening as possible;
- *Maximum tendency to run freely through the blocks in the runner tail circuit.* Suppleness and flexibility of the line or wire rope;
- *Minimum tendency* to twist on itself as load increases;
- *Resistance* to atmospheric agents;
- *Maximum resistance* to wear in the area where the first two turns are taken round the winch drum;
- *Ease* of splicing;
- *Compatibility* of the line with any fixing or recovery mechanisms, such as self-tailing winches, stoppers or (on small boats) various types of cam cleat.

And last – but not least – keep an eye on the cost!

To choose a line or wire rope for the runner tail, the breaking strain of the line (or wire) must be equal to or greater than that of the system as a whole. I want to be particularly clear about this, because it is of fundamental importance not just for the piece of rigging we are talking about now but also applies to everything we will look at later in the field of rigging.

The dimensioning of the tail is not something arbitrary, the result of a debatable decision on the part of the rigger, but – on the contrary – is part of (and must be a part of) a logic of collaboration and respect for the other specialists in the field, which in our case range from the mast maker to the yacht designer. The yacht designer determines the righting moment and as a consequence the mast maker decides the dimensions of the rig; then it is down to the rigger, on the basis of all this, to choose and lay down dimensions and suitable characteristics for the various parts of the rig.

So the rule is that every person responsible for a particular part of the sailing equipment must not introduce an element with resistance less than that specified earlier by the others, so as to avoid a weak link in the chain that would lead to breakdowns and damage. The last element to be introduced must have a breaking strain equal to or greater than that of the other elements already present.

In other words, if we know that the flying block on the runner of a boat has at least – and as a minimum – a breaking load of 3500 kg, which is the breaking strain of the runner cable, and know also that a block that turns a line (or a wire rope) through 180° halves the load on each segment (so each segment of this line must have a minimum breaking strain of 1750 kg), we could experiment with a high-strength double braid polyester line such as Samson XLS ½ inch (12.7 mm) in diameter, with a breaking strain of 8300 lbs (3764 kg).

The reason we err on the high side with the breaking strain has to do with the fact that a line, and also a wire rope, that undergoes splicing and is destined to pass round sheaves, loses some of its breaking strain. In addition, the diameter of 12.7 mm is suitable for the jaws of self-tailing[29] winches.

Having satisfied the first point of the requirements, let us now see if we have satisfied the second. We know the working load on the block is about 2200 kg, which means 1100 kg on each section of the runner tail. We know that – in the best of cases: new line of an excellent brand – a 12.7 mm polyester line at 30% of its breaking strain stretches by 4 to 6%. Thirty per cent of the breaking strain (3764 kg) of our line is 1129 kg: a value very close to the real working load of the tail. The length of the section of tail that is tensioned is about 7 m: 6% of stretch in 7 m means 420 mm: almost half a metre!

You did not expect that, did you? It means, for example, that every time the boat hits a gust, or plunges into a wave, or even worse, does both at once, the runner will be under maximum load and your polyester tail will stretch by half a metre, making mast and mainsail oscillate in a bow-to-stern pendulum movement and also losing tension in the forestay.

Let us try to reduce this stretch by using wire rope for the part under tension and a line spliced to this to use for recovering and easing the tail. The wire rope we need is 5 mm in diameter, in galvanised steel.,[30] obviously flexible, with 133 strands and a breaking strain of 2200 kg.

[29] 'Self-tailing' winches have toothed jaws at the top of the drum in which the final turn of line can be jammed. Thus the handle can be used to turn the winch while the self-tailing jaws recover the line. The self-tailing jaws of each winch, however, will only accept a very limited range of line diameters.

[30] 'Atlantico' wire rope was for many years the solution *par excellence* for running rigging on racing boats, though it has been totally forgotten today, and was not without its defects. It was not made of stainless but of galvanised steel, so it was subject to rust, but it did have some remarkable advantages, such as a much greater flexibility and suppleness than the stainless steel kind, and this meant it was very adaptable in extreme working conditions, around sheaves that were really too small and under high working loads. Also it was not infrequently found on board in the 6 × 19 version instead of the 7 × 19 version used for running rigging. The 6 × 19 was a supple line used for halyards and its core was not in steel like that of the 7 × 19 but in fibre, which made it even more supple though it did slightly reduce the breaking strain. This suppleness had the tremendous advantage of provoking far fewer broken strands than you get with stainless steel wire, thus avoiding those very irritating metallic spikes that do a lot of damage to hands and sails alike. The rust, too, was 'bearable' if you had a good bit of seamanship: Angelo, Gardini's historic mariner, looked after the halyards of the various ocean-going Moro with a mixture of linseed oil and petroleum, which made them last a very long time. Today the pace of life is such that we have no time for such 'spiritual exercises' and so we pay the price (in financial terms because exotic fibres are very expensive, and in ethical terms because with the death of the concept of maintenance a piece of civilisation has been lost) of using materials that require no maintenance.

At our known working load we have a stretch of 44 mm, so we have managed a big improvement on the previous situation: now the elasticity of the line is only a tenth of what it was before! But beware, for what you gain on the swings you lose on the roundabouts! And in fact our wire rope has the following, not to be ignored defects:

- It is very prone to breaking strands, and their loose ends are vicious nails that will attack your hands, your sails and anything else they can reach.
- It is very prone to twisting, and this gets worse as the load increases. And that means that the flying block – if the distance between the eye on the deck and the deck block is not great enough – will twist around itself.
- It rusts quickly (stainless steel does not have this defect but does have, and with a vengeance, the two mentioned above, and in addition has a lower breaking strain than 'atlantico').
- It is heavy; and it absolutely has to have special sheaves, both in shape and diameter.

We are not satisfied with this option, so let us look at another possibility, a low stretch line such as the exotic[31] lines with a core in Kevlar[32] or Spectra[33].

[31] When we speak of line in these pages we always mean good-quality line, which comes from reels bearing the name of the maker, and that maker publishes in leaflets and catalogues the specifications of the line, including its breaking strain. We take no account of anonymous cordage. Worthy of mention in the USA are Samson, Yale and New England Ropes; in England, Marlow and Marina Ropes; in Germany, Gleistein; in Scandinavia, Roblon; in France, Cousin and Lancelin; in Italy, Gottifredi & Maffioli; in New Zealand, Southern Ocean Ropes Also known as Kinnears and Donaghys.

[32] Kevlar, first produced by the multinational Du Pont, and later – in a similar form – and the Japanese Tejin with the name technora, is an aramidic, high modulus fibre in which the material that makes up the fibre is a long synthetic-aromatic polyamide chain in which at least 5% of the amide bonds are directly correlated with the aromatic rings. It has an excellent weight – breaking strain ratio. It has excellent resistance to high temperatures (it burns at 500°C). It has minimal stretch, but little resistance to abrasion, and is also prone to axial compression fatigue. It also loses its mechanical characteristics with exposure to solar radiation. Kevlar, Twaron and Technora are all registered trade marks. Today there exist increasingly sophisticated types of Kevlar, including DSK 75, DSK 78 and Dynex.

[33] Spectra, developed by the American company Allied in joint-venture with the Dutch *DSM* and the Japanese *Toyobo*, was marketed in Europe when it was first launched under the name Dyneema, while today it is also found with the name Plasma. It is a high modulus fibre in polyethylene, the material used for plastic bags but ennobled by a special chemical process called UHWPE. It has an excellent weight-breaking strain ratio; weight for weight it is six times stronger than steel. It has the highest abrasion resistance of any man-made fibre. It has excellent resistance to bending, but very little resistance to heat. It is susceptible to stretch beyond that declared in its specification when it is put under intermittent loads; this atypical kind of stretching is called 'creep'. It is advisable to 'break it in' with at least 50 cycles at 20% of its breaking strain before using it. Spectra, Dyneema and Plasma are all registered trade marks.

A 12.7 mm Kevlar line has a breaking strain of 6350 kg. At a working load of 1100 kg, it stretches by 0.75%, which over the 7 m under tension means a lengthening of 52.5 mm, 8.5 mm more than the wire rope solution. However, we can tolerate this stretch if we bear in mind the advantages we will have compared to the wire rope. Kevlar[34] would be perfect if it did not have the terrible disadvantage of not tolerating solar radiation: and what is more exposed to the sun than your runners?

Kevlar's susceptibility to ultraviolet rays means it has a very short life, and moreover gives no warning when its life is coming to an end. In addition, Kevlar is also particularly susceptible to any curving, including that around sheaves. That is why Kevlar lines tend to break without warning.

At this price level the last resource is to use a line in Spectra.[35] A 12 mm Superbraid from Southern Ocean Ropes has a breaking strain[36] of 5530 kg and at a working load of 1100 kg stretches by 0.62% or 43.4 mm: less than the wire rope, if not by much.[37]

From a strictly mechanical point of view, we could get better results by using Vectran[38] for the runner tails. Vectran stretches even less than Spectra; it has an excellent breaking strain; resists curving fatigue well, much better than Kevlar, and has good resistance to abrasion. However, on the latter we must point out that if you cut open a Vectran line that has undergone a lot of use and pull out the core,

[34] Speaking again of the reliability of rope makers, it is obvious that when we speak of line in Kevlar, Spectra or other exotic materials, we mean line with a core made of as much of these materials as possible. When Kevlar first came out there were ropes sold as Kevlar that had only a very tiny proportion of Kevlar in their cores. And these meant performance well inferior to that of the best ropes in the category. Remember you get what you pay for: lines that cost less than others give poorer performance.

[35] Another matter is that of exotic line with a mixed core, the so-called *blend braids*. Makers such as New England Ropes and Maffioli favoured these hybrid cores: half in Kevlar and half in Spectra, for example. My personal opinion is that blend braids have all the defects of the materials they are made from without these being compensated for by advantages stemming from the combination of the two materials. So I far prefer non-hybrid solutions.

[36] The breaking strains quoted are accurate so long as the lines are spliced by a method approved both by experience and by the manuals. Knots on a line – any knot, on any line – makes that line lose from 40 to 70% of its breaking strain!

[37] Creep – also called cold flow – is stretch caused by deformation and molecular sliding and is particularly common in the family of ropes in Spectra (or Dyneema, or Plasma). So the stretch data provided must anyway take account of this extra stretch that is typical of Spectra and similar fibres, and this has led to these fibres being used more rarely for runners on top racing boats.

[38] Vectran, a Hoechst registered trade mark, is the commercial name of a high modulus fibre called LCP, which stands for Liquid Crystal Polyester, a polyester treated with a special thermal process that enhances its mechanical characteristics. It does not suffer cold flow.

you will see a cloud of particles pour out that say a lot about how much chafing has gone on between the fibres in the core, and how much it has been weakened as a result.

If we want to take a further step forward, first and foremost in terms of cost which here reaches its peak, we could use PBO,[39] a fibre with incredibly high breaking strain: on maxi-yachts 8 mm halyards have been successfully used! But obviously both because of its cost and the particular nature of the fibre, which demands special surveillance during the entire manufacturing process, and because it suffers under abrasion and solar radiation, it is destined for top racing boats.

To complete this picture of the runner tails we cannot avoid mentioning a problem that was a big one in the past, but has recently been solved, though in a way that is not painless for our wallet! I am talking about the age-old problem of fractional rigs which, with the vertiginous increase in the load on the forestay, were literally burning up their runner tails, even on the first outing in a blow! It all happens very simply and very quickly: the first time you bear away to hoist the chute, and thus rightly want to reduce the load on the forestay by easing the runner (the same happens when you get ready to go about), you ease the runner tail – letting it slip out round the winch drum – and the load is so huge that the grip on the drum, the part that is sandblasted to create grip for the sheet, literally melts the polyester. It looked as though a hot iron had been passed over the line, breaking as a result the outer sleeve over the full length that had run round the winch drum.

In the early days this was got round by using wet sandpaper to smooth down the grip on the drum, but obviously paying the price of needing to be more careful handling the runners, since more turns were needed before the tail could be blocked on the winch. Attempts were also made to fit a protective sleeve about 1 m long on the stretch of the runner tail that would sit on the winch under maximum load.

More recently Yale, an American rope maker that is always state of the art, has found a solution using Yaletail, a strange mixture, yellow and black in colour with a Z-shaped weave, made of Kevlar and polyester.

Du Pont Type 77, though it makes the line much stiffer, allows the runners even of an America's Cup class to be eased without risking burning the line.

[39] PBO or Zylon, a high modulus fibre made up of a rigid chain of molecules, has the highest strength-weight ratio of any man-made fibre, and has very little stretch. On the other hand, it has little resistance to abrasion and must be protected from solar radiation. It is very little prone to cold flow.

Table 1.6 Runner tails of equal breaking strain compared

Material	Stretch	Handling	Life	Weight	Cost
Polyester	very high	average	average	low	low
Wire rope and fibre	low	poor	low	high	low
Kevlar	low	poor	very low	low	average
Dyneema Spectra	low	excellent	high	very low	average to high
Vectran	very low	good	high	low	high
PBO	extremely low	good	high	low	very high

To calculate how much line you need for the runner tails, you must multiply the number of line segments in your tackle by the distance between the point where the runner tail is made fast in the cockpit and the position of the flying block[40] when the boom is run out as far as it will go.

If you want to use different materials in your runner tail for the working part and the part that serves only for the initial hauling in, tension the runner and measure the distance between the flying block and the block (or blocks) on the deck, multiply this distance by the number of line segments in the tackle and add the distance between the tackle and the tensioning system (the winch or whatever). If you are using a winch, allow some extra for the turns you will need to take around the winch drum. This section needs to be in line that can stand the working load, while the remainder of the total distance, measured with the runner eased, can be in light line.[41]

One thing to avoid with runner tails is the habit of removing the sleeve from the section of them that runs through the blocks, because this causes in Spectra a problematic tendency to twist every time the runner is tensioned. And this induces very annoying twists in the blocks in the system.

Another thing to avoid is using in the runner tail circuit blocks with flat-sectioned sheaves, like those used for Kevlar, because this makes the line work only on one side of the sheave, putting anomalous strain on it that frequently leads to early and excessive wear on the side bearings.

[40] The flying block should fly as low as possible! This is to avoid an excessively long tail. General usage is to have this block a hand's width lower than the position of the boom when sailing hard on the wind.

[41] It is common practice to splice a smaller line to the end of the runner tail, to avoid having to run out the full length of a larger line, which would be heavier, more costly and harder to feed through the blocks. The disadvantage is that when the runner tail itself is worn, you cannot simple invert it as is usually done because it ends in a light line that would not support the load of the runners.

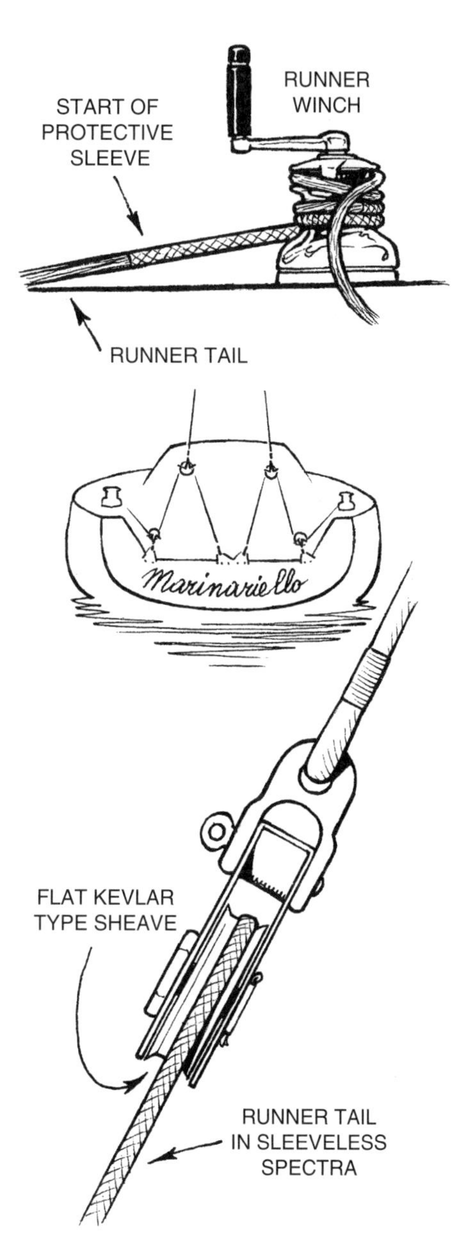

1.18

When lines with a core in Kevlar were very popular, blocks with flat-section sheaves were rightly used so that the Kevlar fibres could work in as homogeneous a way as possible and in particular to make the path taken by the internal fibres – those passing over the sheave – which is shorter, as similar as possible to the path taken by the external fibre, which, since it is longer, obviously creates imbalance. It is wrong to use this kind of sheave with lines not in *Kevlar*. We have had problems of worn bearings on the flying blocks of *Marinariello*, the first IMS Farr 40 seen in Europe, because the Spectra runner tails, which were very small in diameter, always rested on the forward side of the sheave, and as a result the bearing in the block was only under load on one side. To sum up: use sheaves with a deep 'V' section for wire rope, and sheaves with a rounded 'V' section for all lines except those in Kevlar, which require sheaves with a flat section.

Lower Runners

Adjustment systems and types of connection between blocks and upper and lower runners

The lower runners can be adjusted[42] with the following systems when you need – as you always do with fractional rigs – to modify the curvature of the mast: a small purchase with a jam cleat; a block and tackle; a block and tackle or hydraulic cylinder at the foot of the mast; a strop that starts from the mast and deflects the lower runner and is regulated from the cockpit or cabin roof. Basically you need to choose whether to act on the runners close to the flying block or where they are fixed to the mast.

In Fig. 1.19 a simple block and tackle has been opted for to tension the lower runner. The original system was replaced by the block and tackle because it was felt the lower runner would require no more attention once it had been tensioned. A variation on this system is shown in Fig. 1.20 where the purchase is between the runner tail and the upper runner, and an eye splice at the extremity of the lower runner slides along the upper one. This system has the advantage of not making the lower runner twist round the upper one, which is something that often happens with the torsion that builds up under load.

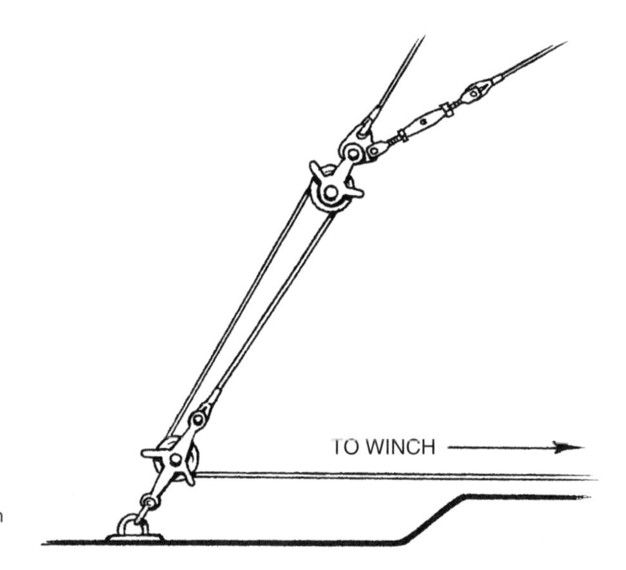

1.19
Here a rigging screw is used to tension the lower runner.

[42] On a masthead rigged cruising boat, with upper and lower runners, it is better to fit the simplest and cleanest system, without any tackle or purchase on the lower runner once it has reached its proper tension. Thus the lower runner will be fixed directly to the upper block of the runner tail.

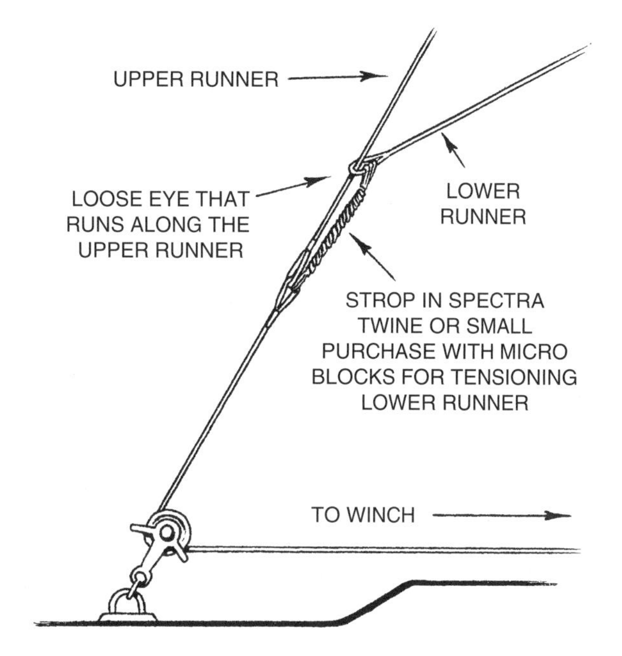

1.20
Tackle for tensioning the lower runner: there is a lashing between the lower runner tail and the lower runner and a spliced eye.

If instead we adopt the system of tensioning the lower runners at the mast, we have the advantage of being able to work from the cockpit and do not have to go and work – moreover standing up – close to the runners.

This system can work in two ways. First, we can run the lower runner down inside the mast over an incorporated sheave and have it come out below deck where there is a purchase or hydraulic cylinder. The second way is to have a wire or light line that takes the same route but is fixed to the lower runner about a metre below where this runner is fixed to the mast. When this line is tensioned it pulls on the runner which is in its turn tensioned.

The first method has been gradually abandoned because it exerts a very heavy load for the dimensions of the sheave built into the mast, for every mast maker will want to make this sheave very small, and it was not uncommon for the sheave not to rotate, or for wires to break threads or lines wear out in a very short time. The second system is more popular today.

If the person choosing the size of the blocks talks to the boat designer and the mast maker on the day of the launch, and on the day of the first tests on the water, he will not witness the umpteenth (unsuccessful!) attempt to weld the terminal of the upper runner to the pin of the block and both of these to the lower runner.

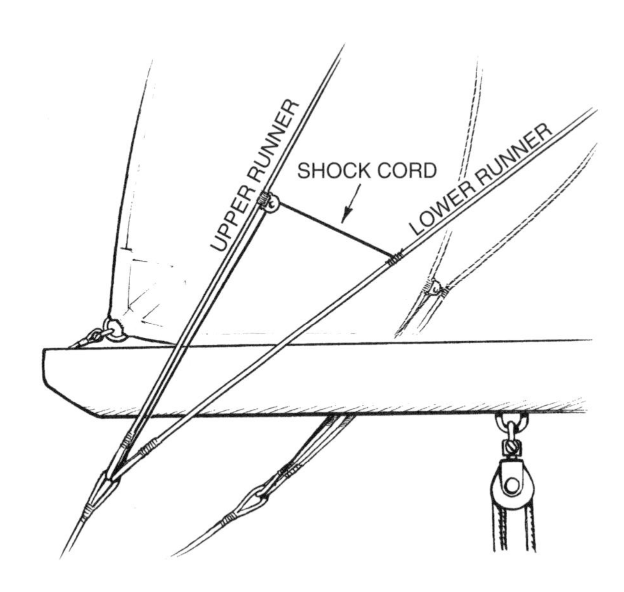

1.21

How the lower runner is brought alongside the upper one by means of a shock cord so that when going about it does not tangle with the boom.

The secret is quite simply (think first!) to ensure that all those involved in the operations of dimensioning the flying block and the runner line terminals keep well in mind that these three elements must be compatible!

Mast Attachments of Upper and Lower Runners

The easiest, most immediate and economical way of fixing the runners to the mast is undoubtedly to use 'T' terminals[43], as shown below.

To fit this system, the mast maker or rigger must drill a hole in the side of the mast and insert a specially made plate to house the 'T' terminal. This plate is often held in place with rivets.[44] The load borne by the plate is passed to the mast, as it is specially shaped to make this happen naturally.

[43] Gibb has for some time been offering a high load version of the classic 'T' terminal which costs more but is highly recommended, especially for the runners on a fractional rig.

[44] It is preferable to use high-quality monel type rivets in steel – not aluminium ones! – although they need a particularly powerful gun. I have had to install lower runners with the mast already stepped, and I had to use a diving cylinder to drive a compressed air gun because the monel rivets were too tough for a normal gun. Take care always to insulate steel rivets from the mast with anti-corrosive Duralac paste to avoid corrosion quickly setting in.

1.22
System for attaching runners in steel rod to the mast.

The 'T' terminal is inserted perpendicular to the mast and once inside is rotated so as to point downwards: the weight of the runner will hold it down and thus prevent it coming out. However, it is advisable to insert a rubber plug[45] in the opening that remains above the terminal to rule out any nasty surprises. If a plug is not available you can fit a safety plate instead. Another precaution to take with 'T' terminals is to tie a shock cord around the shank and pass it round the mast, so that when the terminal is at rest (when the runner is eased) it remains always parallel to the mast. Unfortunately it is not uncommon, especially when you hit a wave, to end up with the shank pointing upwards towards the masthead, and if you do not notice it is in this anomalous position and tension the runner, the effect is like cutting through the runner, with all the consequences that entails. It is highly advisable to inspect the terminal frequently. In some fractional rigs an eye terminal inserted directly into the mast has been used, with the runner fitted directly to it and leading down to the runner tail. This is definitely a solution to be discouraged, for it is very unlikely that an eye positioned in this way offers the freedom of movement required for this application.

[45] Gibb was the first to provide plates for 'T' terminals with this special protective plug.

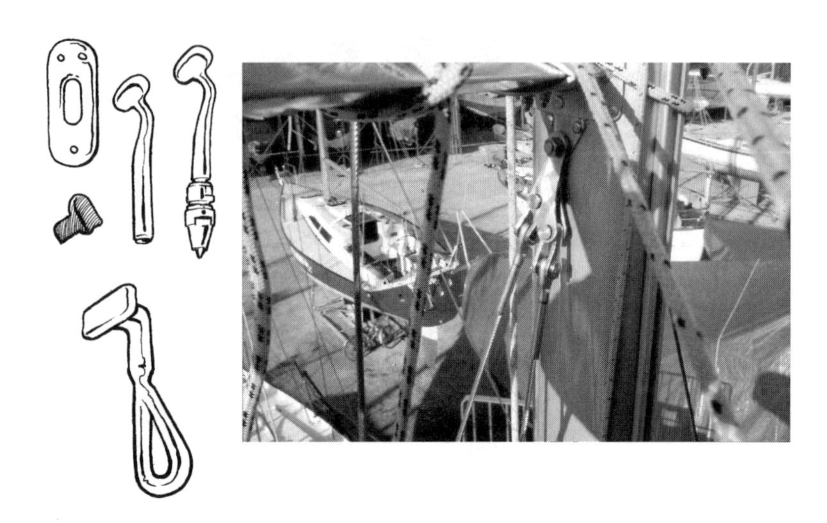

1.23

Top left: high load Gibb 'T' terminal (note the rounded head); if the rubber plug that blocks the terminal is not available a safety plate can be used; bottom left: special 'T' terminal for use when passing from traditional wire rope runners to exotic ones with single braid line such as Spectron 12 or normal double braid; right: runner attachment as used in less recent times, certainly heavier but decidedly more reliable. A pin passing through the mast held a couple of chainplates for an eye terminal.

In the photographs above we can see an attachment for the runners traditionally used on less recent boats: it is certainly heavier but decidedly more reliable. It is essentially a pin with a pair of chainplates attached for fitting the eye terminal. Since the chainplates can swing in a stem-stern direction, the runner can be correctly angled when under tension, while the way the plates are made allows them to follow the correct transverse alignment simply by bending slightly. On the fleet of the 2001–2 round the world Volvo Ocean Race we saw runners fixed to the mast using chainplates similar to the traditional ones but with several turns of Spectra between the plates and the runner terminal. Something of a return to the past, though with modern materials.

The Backstay

Although the backstay is present in fractional rigs, on these boats it is considered somewhat as a secondary part of the rigging since the mast is held upright by the runners, which play the major role in determining mast bend and hence sail shape.

The opposite is true in masthead rigs, where the backstay holds up the mast and plays a structural role, while the runners have a secondary function, as we have already described.

In a masthead rig the backstay is essential for determining the tension of the forestay, and thus the shape of the genoa, and also the shape of the mainsail. When the sea is choppy in relation to wind strength, the backstay is eased just enough to fill out the genoa and develop maximum power to allow the boat to cut easily through the waves. With fresh wind and calmer seas the backstay is tensioned to reduce forestay sag, thus flattening the genoa and allowing the boat to point higher.

On 3/4 fractional rigs, at least those of up to 50 feet overall, a purchase, usually 6:1 and mounted between the backstay and a chainplate on the deck, is sufficient for tensioning the backstay.

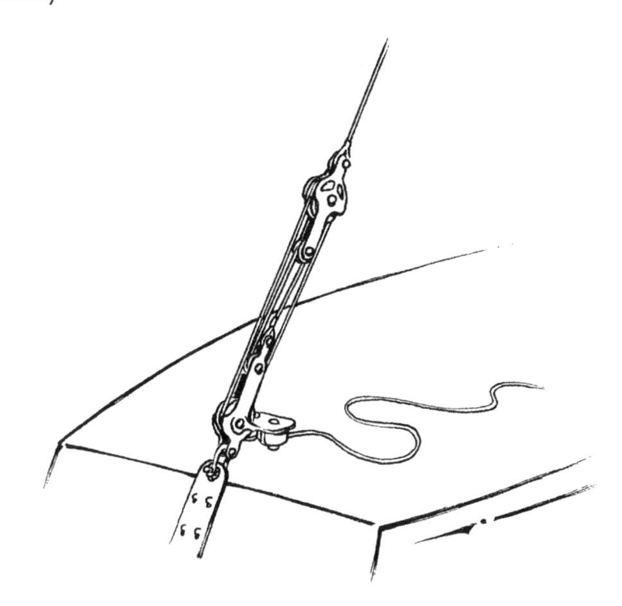

1.24
4:1 purchase between the backstay and the chainplate on the hull.

The purchase on fractional rigs is both light and simple. Obviously you cannot expect it to provide very high tension because of the intrinsic limitations of a purchase system. It is suitable for fractional rigs that need little tension, since the backstay has only to bend the very top part of the mast which is short and thin and thus easy to bend even with low loads applied.

On all rigs with a split backstay or twin backstays, from J24s to fast cruising maxis, the backstay divides into two at a certain point forming an inverted 'Y', one end of which is fixed to the port side of the transom and the other to the starboard side. As for example on the J24, a pincher block is fitted to each leg and from this a purchase is fixed to the transom and tensions both, thus tensioning the backstay. On bigger boats, from those in charter fleets to fast cruising maxis such as the Wally, the upper part of the backstay ends with a block through which a line or cable runs, and tensioning this line puts tension on the backstay.

You may be wondering why it was necessary to complicate life on board by fitting a split backstay. The only reason is that it makes life more practical in the transom zone. On the J24, for example, as on all boats where the tiller passes over the transom, the backstay needs to be split, otherwise it would interfere with the movement of the tiller. The only way to avoid splitting the backstay to free up the tiller is to fit the backstay to a hoop under which the tiller can move freely. But this can only be done on boats where the load on the backstay is small, because otherwise for mechanical reasons this hoop would have to be made so massive that it would be unacceptably heavy and cumbersome. The illustration below makes it clear how a split backstay is tensioned.

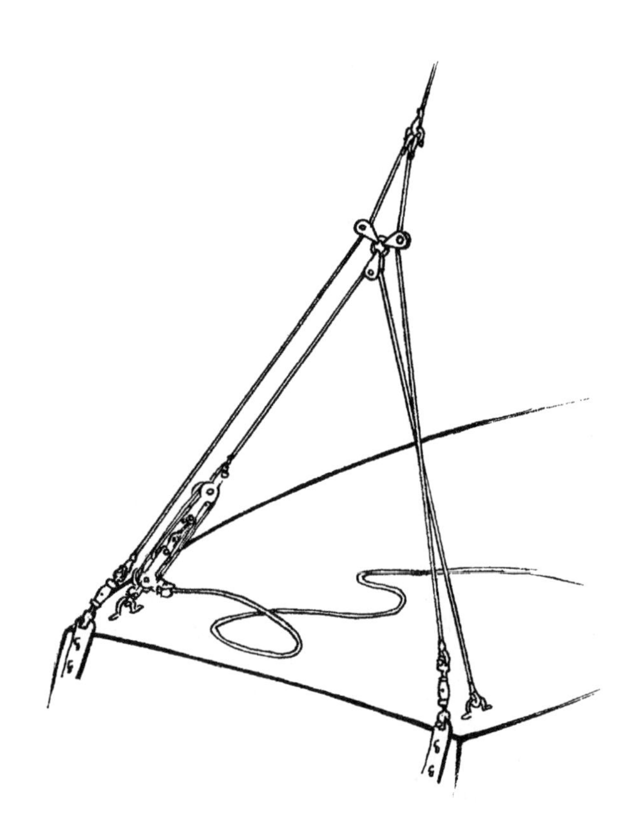

1.25
Split backstay in the form of an inverted 'Y'.

A further essential advantage of this system is that it is very reliable. Even if the tensioning tackle breaks, the safety of the mast would not be in any danger because the fixing of the backstay is independent of the tensioning system.

The greater the internal angle between the two lower arms of the 'Y', the more efficient this system is. Tensioning is highly efficient when you start tensioning with the backstay loose. The efficiency decreases as the two arms become almost parallel, the point where tension on them reaches a maximum. Here the breaking load is not so important, as it was in the previous examples, because even if the purchase breaks it does not prejudice the safety of the system.

Table 1.7

Length overall of boat in feet	purchase of tackle
24	4:1
26	6:1
28	8:1
30	12:1

Above is a table that allows you to choose the right purchase, which is the correct reduction of the force required, according to the length of the boat in feet.

Actually the fact that this system is so popular on *J24*s reveals another of its basic advantages, which is that it permits extremely fast regulation. In fact, on cruiser-racer boats such as the C&C 37′, a masthead rig with a double[46] backstay (a solution that makes life much more comfortable when cruising because you can disembark over the transom or use a gangway), the same system was used to tension the backstays.

On pure cruising boats such as charter boats, or fast cruisers such as the Wally or the latest Swans, the reason for having a split or double backstay lies in the ease of access to the stern platform, which is often a hinged part of the transom that is folded out. The Wally in particular needs a split backstay because normally it has a locker for the tender at the stern.

When using a backstay system with a pincher block and a flexible cable for tensioning, you must pay particular attention to the following factors: the choice of the block in relation to the cable or line used, and the choice of length of the backstay tensioning tackle. Let us take a detailed look.

[46] The split backstay consists of a single wire rope that splits into two at a certain point, while the double backstay has two independent wire ropes.

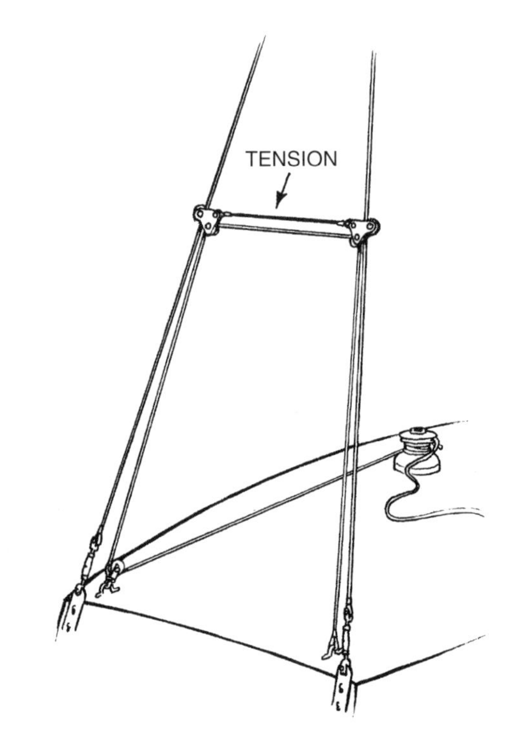

1.26

System used on the C&C 37 to tension the double backstay.

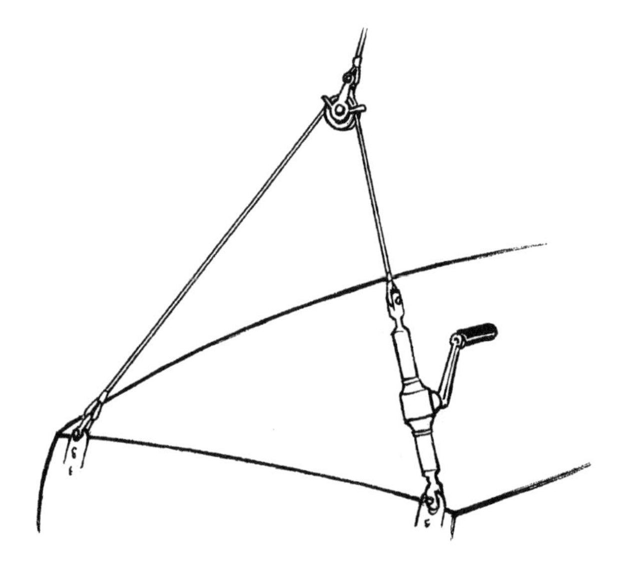

1.27

An inverted 'Y' split backstay with a tensioning system consisting of a flexible wire rope that passes round a block and is fitted to a mechanical tensioner with handle.

When choosing blocks, to begin with you must use blocks with a breaking load equal or greater than that of the backstay cable, which means using a rather large block. Then you must consider that you need a block with a sheave suitable for the cable or line that will run through it. If you use a metal cable the section of the sheave must be a deep 'V' so that the cable can lie well inside it. And apart from the question of the section, bear in mind that every cable, because of its diameter and the way it is made, does need a pre-established sheave diameter.

Here is an example. A flexible cable of the kind used for halyards, made up of 133 wires with an external diameter of 6 mm needs a sheave diameter of 150 millimetres.[47]

Generally for line, the recommended sheave diameter is 8 times the diameter of the line, except for Kevlar line that requires a sheave diameter 12 times its own. As far as sheave width is concerned, it should be 10% more than line diameter.

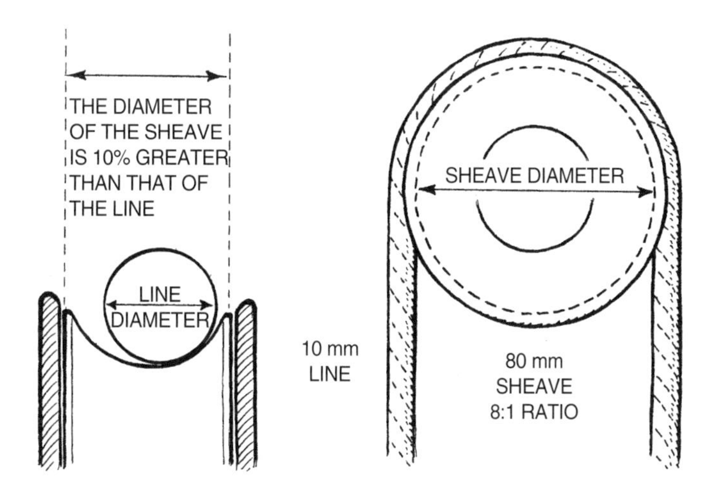

1.28
Indication of the diameter of the line and of the sheave (Sampson).

Using line simplifies everything a lot. The sheave can be the classic one found in the big majority of blocks, and the diameter of the sheave dictated by the diameter of the line is never as cumbersome as when using cable.

Another important factor to bear in mind, especially for big boats of more than 60 feet, are the bearings of the block.

The backstay block will never have large lengths of line or cable running through it, as happens with halyards or sheets, so its ability to let long lengths of line or cable run through with low friction is irrelevant. On the other hand, what does matter is its ability to sustain high loads with little line running through. This job is done perfectly by a block with axial load bearings in a composite race.

[47] The diameter of the sheave is obtained by multiplying the diameter of the line by 25.

A practical note. When deciding the length of tackle, many people make the mistake of using the same tackle on a split backstay that they used on a single backstay. They forget that when the backstay is split you need double the length to get the same degree of tension you had with a single backstay.

Frequently used on masthead rigs with a single backstay are a simple rigging screw fitted on the stay and a mechanical tensioner between the stay and the deck actioned by a winch handle. A more sophisticated alternative is a hydraulic cylinder.

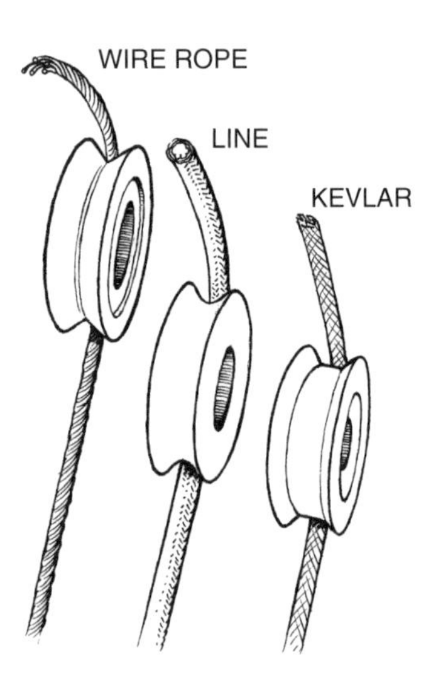

1.29
Sheaves for different uses.

The classic solution for masthead rigs with a single backstay is a rigging screw[48] fitted on the stay. Once adjusted to trim the mast it remains as it is until the next adjustment.

A decidedly more sophisticated tensioning system uses a hydraulic cylinder. This may be integral, in the sense that a handle mounted on the cylinder itself activates the hydraulic pump inside, or with a hydraulic tube leading to a central unit that can also provide power to tension the babystay, the outhaul and so on.

To give a good example of the problems of tensioning the backstay I will explain how the owner of a Passatore, a fine design by Jean-Marie Finot, a masthead rigged 34 footer, modified his backstay. He enjoyed using his Passatore

[48] A rigging screw or mechanical tensioner has two rods with opposite threads, one left- and one right-handed, so that when the central body is rotated it pulls in both rods, thus tensioning the wire rope or line that is fixed to it.

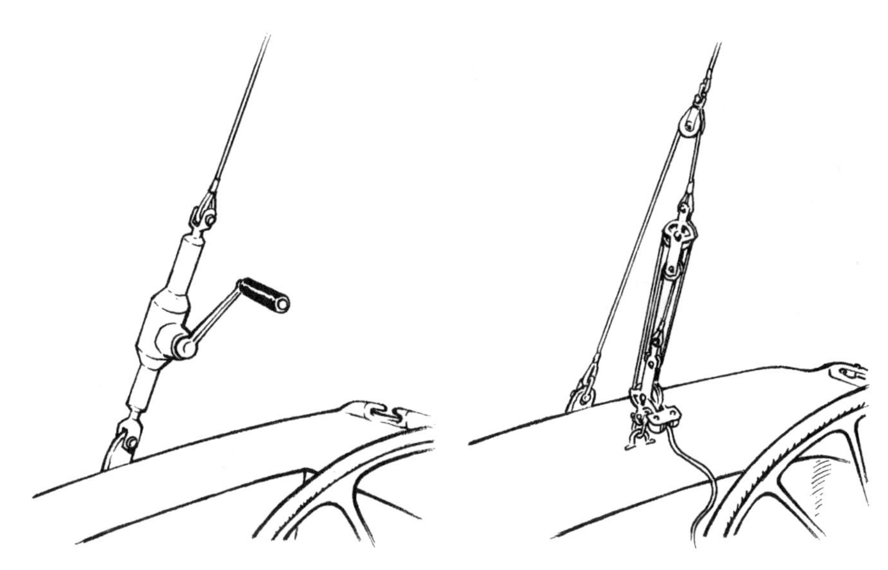

1.30
Mechanical backstay tensioner and, right, a block and tackle system.

not just for the classic summer cruise but also in hard-fought winter club races. So he needed to adjust his backstay tension much faster than he could with his existing system, which was good but slow.

The system he had on board is illustrated here. The datum to start from is the breaking strain of the backstay, which is a rigid spiral cable with 19 threads, a diameter of 7 mm and a breaking strain of 3500 kg. As we have seen in the case of runners, here too we must select a block that can bear a load equal to or greater than the breaking strain of the stay. We could for example choose a Lewmar 2990 2808, with an 80 mm sheave and a breaking load of 5000 kg.

The block divides the load in half, so the upper triple block in the illustration must have a minimum breaking load of 1750 kg, such as the Lewmar 2990 1603, with a sheave of 60 mm and a breaking load of 2000 kg. Below we need a triple block with a becket and cam cleat, again with a minimum breaking load of 1750 kg (Lewmar 2990 1610). In effect we have created a double purchase, known as a cascade. In this a first 2 : 1 purchase leads to another tackle with 6 : 1 purchase, so to calculate the combined purchase we need to multiply the two together and so arrive at 12 : 1.

At this point the safety criterion is satisfied by the correct dimensions of the blocks, but is the system as efficient as the earlier one? In other words, will our skipper be able to tension the backstay more quickly than before and with the same tension as before?

1.31
Harken blocks in cascade on a Vallicelli Genesi.

The answer is partly negative and partly positive. When low loads are required the system is decidedly faster than the original one, but if slightly higher loads are demanded this 12 : 1 purchase just is not up to it! If we consider that someone who does not habitually sail 300 days a year is able to haul, while standing in the boat, about 34 kg, multiplying this by 12, which is the purchase available from the tackle, the theoretical load is 408 kg, not considering loss of efficiency through friction. And 408 kg are not a lot compared to the working load of the backstay, which with a breaking strain of 3500 kg generally will have a working load equal to 20% of the breaking strain, that is 700 kg. So let us try increasing the purchase of the tackle by inserting another single block with a breaking load of 1750 kg after the first *Lewmar* 2990 2808, which will lead to a triple block like the one used in the previous example but with a lower breaking load of 875 kg. The overall purchase is now 24 : 1, giving double the previous load at 816 kg, in line with the 700 kg working load of the backstay.

Genoa sheets: the forces involved

In John Bertrand's book *Born to Win*[1] there is a photograph of the author with the genoa sheets of *Gretel*, a 12 m class that was the Australian challenger in the America's Cup. It dates back to an era in which the running rigging on a racing boat had to be in steel wire spliced on to a fibre line.

Today, even on a maxi yacht, the genoa sheets are all in fibre, thanks to the introduction of lines with a core; in particular high-resistance exotic fibres. Still that photograph is striking evidence of the forces involved when we use the sheet to trim a genoa. If an ordinary polyester line would have sufficed for the genoa sheet – in terms of stretch, breaking strain and practicality – you can be sure that everyone would have been only too happy to avoid using wire rope, even in galvanised steel.

But what are the forces involved? How can we understand what line to use for the genoa sheets? At first we might be tempted to use the same reasoning we used for the runners, and base our thinking on the breaking strain of what is at the other end of the sheet: the cringle on the clew of the sail. If this cringle breaks at 1000 kg, we could use a line with a breaking strain equal to, or preferably greater than, this value.

This line of reasoning would not be entirely wrong except that, unfortunately it leads to us having a sheet that is terrifyingly over-dimensioned. The clews of sails, with all their reinforcements in fabric, and the breaking strain of the cringles pressed or stitched into them, have strength that is out of all proportion to the effective working strain on the clew. So starting from the effective breaking strain of the clew of the sail would lead us to use a line with a huge diameter, bringing with it problems of cost and weight and being generally impractical.[2] So we need

[1] John Bertrand, *Born to Win*, Sidgwick and Jackson, London.

[2] Remember you must reeve the line through any deck blocks and fit it into the jaws of a self-tailing winch, if used.

a realistic starting point, and this could be a tension gauge applied to the clew, though this is obviously only practical on America's Cup class boats, maxi yachts and the like – or a theoretical value.

And here there is a fairly simple formula to help. It may not have been invented by naval architect Roger Marshall of the Sparkman & Stephens studio, but he certainly made it popular. This allows us to calculate the working load on a sail such as the genoa as a function of sail area and (note well!) the square of the apparent wind strength, all multiplied by a coefficient. So the working load in kilogrammes on the clew of a genoa is equal to:

$$SA \times V^2 \times 0.02104$$

where SA is the sail area of the genoa in square metres, V is the apparent wind speed in knots and 0.02104 is a coefficient. If we want the load in pounds,[3] we put the sail area in square feet[4] and use the coefficient 0.00431.

I want to underline both the unique value of this formula, which can give us a practical idea of the loads involved, and its limitations in real life, for it does not take account of the displacement of the boat.[5] However, it is advisable to use the Marshall formula for boats of average displacement: it would not be a valid way of calculating the loads on the genoa of a multihull or of a ULDB.[6]

In fact, in all cases of hulls with very high or very low displacement, the values given by the formula need to be corrected respectively upwards or downwards, by a factor that only the experience of the rigger can determine. But, I repeat, in most cases this formula yields realistic and relatively accurate values.[7]

[3] To convert pounds rapidly into kilos without a calculator, simply divide by two. For greater precision, multiply pounds by 0.4535 to obtain kilos. Numbers representing quantities in pounds are often used on rigging hardware to indicate loads. For example, the Sparcraft 10 shackle has a working load of 10000 lbs. Nitronic 50 rod size 10 has a breaking load of 10300 lbs. To convert inches to millimetres, multiply by 25.4.

[4] Multiply square feet by 0.0929 to get square metres.

[5] We will see later that this is only partly true.

[6] ULDB stands for Ultra Light Displacement Boat, also known as a Sled. These boats designed in California are narrow in the beam in relation to their length and are very light. They were successfully used in various editions of the Transpac race.

[7] It is very probable that other naval architects, or especially sail designers, have elaborated and used more sophisticated formulae than the one described. It is also known, in this context, that sail lofts use such dedicated programmes as Flow or Membrane, which have to run on machines not everyone can afford. The fact is that still today the Marshall formula remains the only one available to the general public, while the only other field solutions available are the usual 'they say' or 'so-and-so does it like this', and these can hardly be verified.

Let us look at a practical example, on the genoa of an IMX 38. The number one genoa, the largest, is of 51 m^2 while the number three, the 100%, is of 34 m^2.[8]

We have chosen – and it was no accident – these two sail areas in order to demolish one of the myths that still deceive far too many sailors into thinking that a sail of a greater area leads automatically to higher loads, so the number one genoa must give rise to higher loads than the others. This is absolutely not the case.

We can use the number one genoa, when sailing upwind, with up to a maximum of 25 knots of apparent wind, while we can fly the number three with up to 40 knots. These are limit values for the optimal use of the two sails, but we should take them into account because we may easily run into puffs of these strengths before we have time to change the sail, so the genoa sheets must be able to handle the peak loads involved.

Applying the Marshall formula, we find that with 25 knots of wind the number one genoa will have a maximum working load of 669 kg, while at 40 knots the number three will develop a load of 1142 kg! This is because although the number three has a smaller area, the wind strength in which it operates is far greater, and since this factor is squared in the formula it leads to a big increase in the resulting load.

2.1
Wire-fibre splice on a genoa sheet with Barient shackle.

And so, with a working load of 1142 kg, we need to select a sheet with a breaking strain about one and a half times this value to ensure a safety factor[9] of 1.5.

[8] Genoas are denominated as #1, the one with the greatest area, to the smallest, generally the #4 or even the #5 used on maxi yachts, apart from the storm jib. Number one genoas (which may be divided into light, medium and heavy according to the weight of their cloth) are also defined as '150%', while the number three is also called the '100%': these percentage values refer to the 'J', a rating measurement roughly equal to the distance between the forward edge of the mast and the attachment of the forestay on the deck. Thus with a 'J' of 5 m, a 100% genoa will have a foot 5 m long and a 150% genoa one 7.5 m long.

[9] On cruising boats, the safety factor must be at least 2.

This will mean a supple[10] stainless steel wire rope $\frac{7}{32}$ inches (5.5 mm)[11] in diameter, with a breaking strain of 1750 kg. Basically it will always be the number three genoa that dictates the choice of the correct sheet because of the high winds in which it is used.

The wire rope, which we cannot handle directly, will need to be spliced, using a wire-rope splice, to a fibre line which will be easy to handle and will form the remaining part of the sheet. The other end of the wire rope will be fixed using a nico-press to a specially designed shackle called a *j-lock* or *press-lock* that allows the sheet to be fixed to the clew of the genoa.

Barient, an American company once famous for its winches, was the first to sell these shackles, which were called *j-lock* because their shape resembles a letter 'J'.

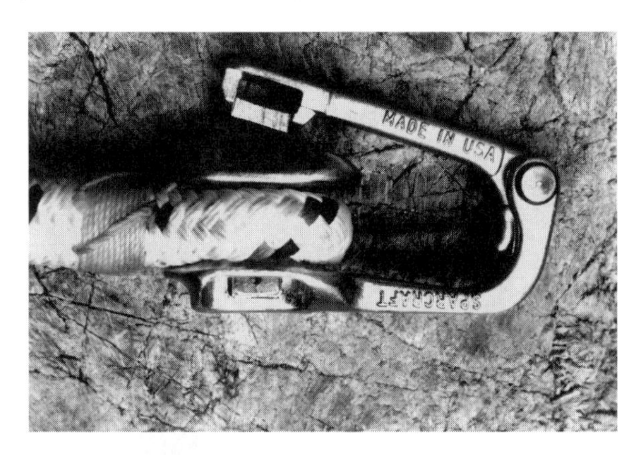

2.2
Sparcraft press-lock shackle.

The big advantage of these shackles, which are very compact, is that the wire rope can be fitted directly to them as their base acts as a thimble, and their compactness means that the clew of the genoa can be trimmed until it comes into contact with the sheave on the genoa track. They are only 3 or at most 4 cm long and this is obviously shorter than the bowline that would be used with a fabric sheet.

[10] Wire rope for marine use is basically divided into three types: 'extra-flexible', 'flexible' and 'non-flexible'. Extra-flexible is suited to pass around sheaves and thus has to be made of many strands, since the fewer strands wire rope has the stiffer it becomes, and these will tend to break if it is bent. The extreme is rod, which cannot even be coiled beyond a certain diameter. Extra-flexible wire, typically used for halyards, is also called 133 strand, or 7 × 19, since it is made up of 7 strands which in turn are each made from 19 wires twisted together. With galvanised steel wire rope, it was common to use a 6 × 19 wire with a core in fabric to guarantee even greater suppleness with minimum loss of breaking load. Non-flexible wire rope with 7 or 19 wires, that is 1 × 7 or 1 × 19, is made of 7 or 19 equal wires and usually used for shrouds and lifelines. 49 (7 × 7) flexible wire has 7 strands each of 7 wires and is a half-way house.

[11] Again, don't panic! To convert $\frac{7}{32}$ inches to millimetres, divide 25.4 by 32 and multiply the result by 7. Easy, isn't it?

In addition, *j-lock* shackles can pass through the swallow of a modern genoa track even with a sheet in tension already there, and this is very useful when preparing a sail change on the same tack.

The only weak point of the *j-lock* was the pin that held it closed, for if this was subjected to knocks, especially when going about when the shackles often beat against the mast and shrouds, the shackle could spring open. To try to get round this problem, the *j-lock* was bound with *gray-tape*[12] to keep the pin in the safe position.

The problem was solved by a similar shackle made by Sparcraft, the famous American mastmaker: the *presslock*, which took its name from the way it was opened, by pressing two small metal buttons sited in a depression on the body of the shackle. The *presslock*, in a special steel called 17-4-ph,[13] unlike the Barient product which was made from AISI 316 steel, had a very high breaking load and became the standard for genoa sheets on countless racing boats. It was even used as a mini-block for the heads of mainsails that required a 2 : 1 purchase for their halyards.

The only less than excellent aspect was the difficulty of opening the shackle with cold hands when sailing in low temperatures, for the buttons to press, especially on the small *presslock*, the *junior*, were very small indeed.[14]

Another typical use for the *j-lock* and *presslock* was on the flattening reef line of the mainsail, a small line that was tensioned when the wind freshened to eliminate the 'belly' of the sail along the foot, which is made from light material. The line is a rope or a rope spliced to a wire rope that passes inside the boom like the other reef lines and ends up rove to a cringle or eyelet on the leech about half a metre above the clew. It is quite simple to use: the line is tensioned with the vang loose and the mainsheet eased slightly. (Today the *j-lock* is made by the TYLASKA company.)

[12] 3M gray-tape is a waterproof fabric tape that had a lot of success on racing boats as a way of taping, fixing and protecting a whole range of things, such as the ends of the spreaders, the protest flag wrapped round the backstay, the pins of the rigging screws and so on. This tape, which actually originated in the automotive sector, is useful for emergencies but does not last long in the marine environment and tends to crystallise and come apart in the sun, and so has to be continually replaced. Much better is self-amalgamating tape which holds by sticking to itself and not the thing it is wound round, unlike gray tape which leaves an annoying film of glue on surfaces.

[13] Types of ARMCO steel, such as 17-4-ph, 17-7-ph, or even 15-5-ph var, with names ending in 'ph' are obtained by a special fusion technique known as precipitation hardening, which hardens the metal as it is fused to give it better mechanical properties and excellent resistance to oxidation.

[14] I would like to pay homage to an Australian rigger, Peter Gardner, who rigged the Australian 12 Metre *Kookaburra*, and was the first in the world to use a kind of *j-lock* for high loads, which was practically a piece cast in metal in the shape of a shackle that ended with a rounded thimble designed to house the wire rope spliced with a *talurit* swage. This was known as the Gardlock and for many years was the only kind of shackle that could be used for the genoa sheet of a maxi yacht. But the advent of the Sparcraft j-lock maxi marked the end of the Gardlock, because this had the disadvantage of having to be opened or closed with a key.

2.3
Equiplite genoa sheet system.

The latest in genoa shackles is the one made by the Australian company Equiplite that basically uses lashings of exotic fibre between the genoa clew and the sheets, held in place with a protective Velcro fastener. However, to fit a *j-lock* to the genoa clew we will in any case have to use one of the commonest elements in rigging, which dinghies, small keelboats and ocean-going yachts have in common: the nico-press or talurit swage sleeve.

In fact we are not obliged to use this device, as we have a full three alternatives. The first is a splice done by hand to form an eye on the wire rope. The wire rope is unravelled, the strands separated and then woven according to a special procedure and sequence.

The advantages are that this yields a breaking strain that is almost equal to that of the wire rope itself, and there is no need for special presses. But things are not quite that simple, for while it is possible to splice a wire rope by hand using only a small metal marlinspike, it is also necessary, on larger wire ropes, to use a special vice to hold the wire taught around the thimble and thus allow the splice to be done with the marlinspike. This vice is advisable, even on wire ropes of only 8 mm. The disadvantages of hand splicing are the fact that you need a specialist to do it, for these splices take years of practice to master. Another disadvantage is that the wire rope will be thicker along the full length of the splice and this, in the case of halyards, means the thimble cannot be brought up tight against its sheave.

One thing not many people know is that, contrary to what you may think, even a stiff 1 × 19 wire rope can be hand spliced as long as this is done around a thimble of appropriate – very large – dimensions.

The second alternative is a series of 'U'-shaped steel clips with threaded ends, which are passed through a saddle. Two nuts are screwed on to press this saddle against the two lengths of the wire rope and hold them in a 'deadlock'. Obvious advantages are that these clips are easy to find, easy to fit, can be re-used and, last but not least, are very economical.

The disadvantages are: the considerable bulk of the result, for three clips are required for safety reasons; the difficulty of binding the fitting to avoid other lines snagging on it; and the fact that it cannot be used on such lines as halyards that need to be taken in by their full length, since the three clips must occupy a length of not less than 15 cm.

Note that the fixing nuts on the clips must all be on the side of the longer length of wire rope.

The final alternative to the *nico-press* swage is the terminal. This terminal[16] may be pressed or machined from a bar of solid metal. Generally – and preferably – it is made from AISI 316 steel, and can have various shapes, depending on the job it has to do: it will have an eye terminal for use with female chainplates or a fork terminal for male chainplates.

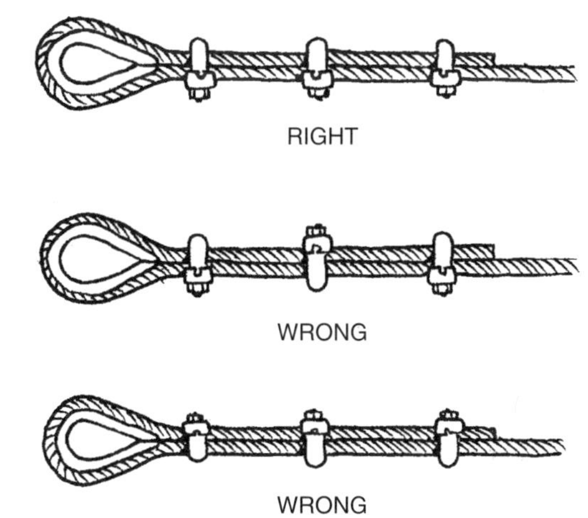

2.4
Correct way to fit clamps.

Big advantages of these terminals are that there is no need for a thimble, and that there is a choice of types of terminal: eye, fork, 'T', hemisphere, etc. The fact that there is no need for a thimble is very important in the case of rigid 19 strand wire rope, where a thimble of the correct diameter would be ridiculously large. On the downside, terminals are costly and require specialised machines and

[15] The cringle is a metal eye crimped into the sail through which a reef line may pass, or it may be used to fix the new tack of the reefed main on to the fixed hook at the gooseneck.
[16] Contrary to what people think, there is nothing against using these shroud terminals on flexible wire, you just have to remember that the breaking load of the splice is that of flexible wire and not rigid wire!

manpower. Looking at the strong – and weak – points of the other systems illustrated, it is clear why this kind of terminal is so common on all kinds of boats and in so many applications.

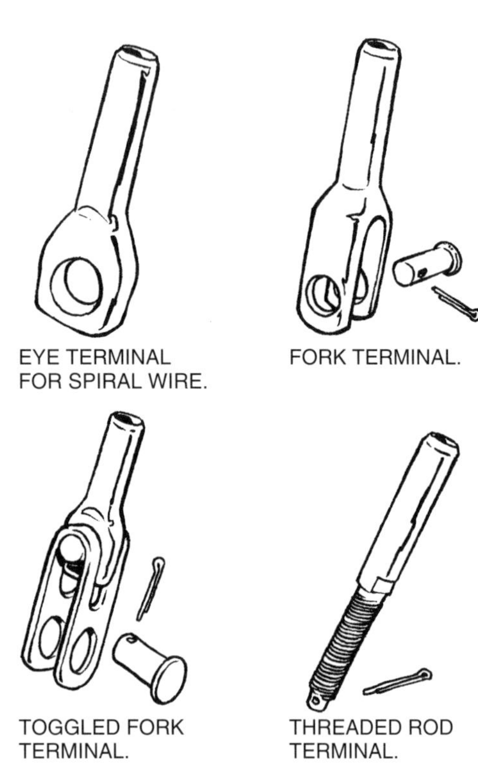

2.5

The *talurit* is a ferrule of German origin with an oval section, as opposed to the figure-eight section of the American nico-press. The difference in the sections is no accident: the talurit needs pressing only once, regardless of the diameter of the cable, with a compression tool whose jaws reduce the original section of the sleeve to the diameter necessary to guarantee the required strength.

2.6

Nico-press swage and, right, talurit.

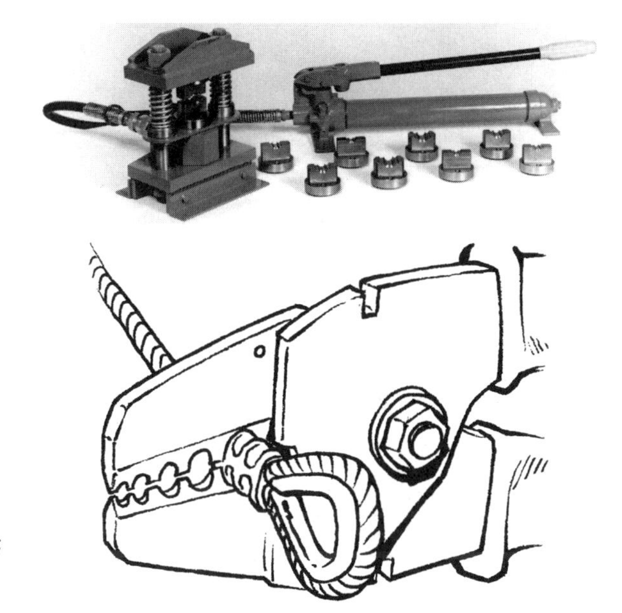

2.7
Hydraulically operated press for talurit; below, a manual press for nico-press swages.

The nico-press demands a different pressing technique with a compression tool – with specially shaped jaws – and a number of pressings, depending on the diameter of the cable.

The illustration on this page shows a compression tool being used on a nico-press, while the drawing on the following page shows the three pressings necessary for a nico-press on a 5 mm cable and the order in which they should be carried out, starting from the central one.

The sleeve should always be crimped first in the centre, then next to the thimble and finally towards the standing part. If the *nico-press* needs more than three pressings, as it will on large diameter cables, the fourth should be effected next to the third, further towards the standing part.

Note that the drawing also shows an important detail: the two extremities of the sleeve must not be crimped, for the cable, both where it enters and exits the swage, must have freedom of movement, for otherwise its strands will break prematurely.

To use both the *nico-press* and the talurit correctly, account must be taken of the fact that, after pressing, the sleeve will lengthen both in the direction of the thimble and towards the standing part, so care must be taken to leave a certain space between the point of the thimble and the end of the sleeve. Otherwise the sleeve, once pressed, will slide up over the end of the thimble and not allow the cable freedom of movement. Also, a small length of cable should be left poking out of the sleeve alongside the standing part, so that it does not disappear inside the sleeve after pressing.

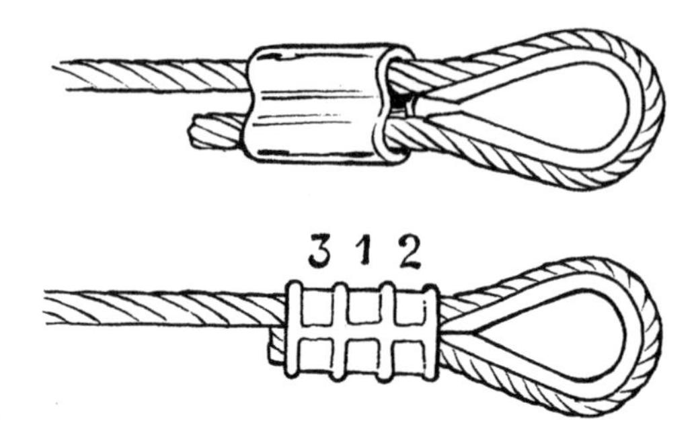

2.8

Nico-press swage showing sequence of pressings.

A practical note. Below is a table that allows you to check whether your nico-press/talurit has reached the reduced diameter that guarantees the strength required.

cable diameter (mm)	number of pressings	final diameter (mm)
3	3	7.9
4	3	10.5
5	3	12.5
6	3	14.0
7	4	16.1
8	5	17.8
10	5	20.5

The left-hand column shows the cable diameter, the right-hand one the diameter you should read on the gauge in the depressions on the pressed sleeve. The centre column shows the number of pressings required on the sleeve (this applies only to the *nico-press*).

In the *nico-press* system, with its typically American pragmatism, one or more templates are provided together with the pincers, providing a test method the Americans call 'idiot-proof': either it's right or it's wrong!

On each template, or on each part of it, if it is for use with different cable diameters, is shown the type of die to use for each cable diameter and alongside the sleeve to use for that cable (the sleeve has the same symbols[17] stamped on it). The slot corresponding to these indications can be used as a gauge on the pressed sleeve: if the sleeve fits inside, it is correctly pressed, otherwise it must be discarded or pressed again.

[17] This is true only for swages of reputable brands, such as the original nico-press and those made by Loos.

In the case of a talurit, after checking the reduced diameter using the table above, it is a good idea to file off the two burrs that form along the body of the sleeve after pressing.

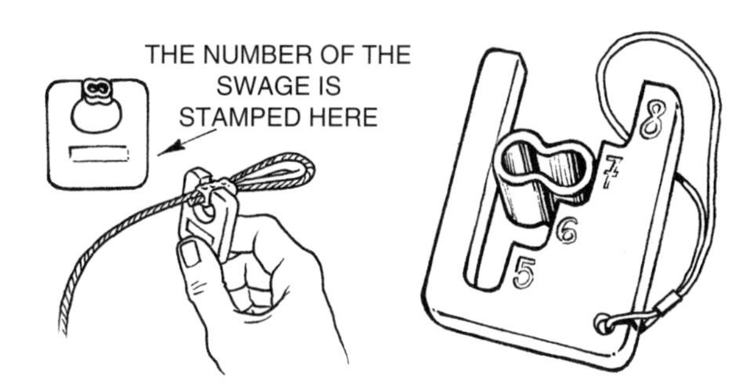

2.9
(Left) Single gauge for checking whether the nico-press has been correctly crimped and (right) a multiple nico-press gauge.

With both systems it is advisable to wind some tape around the short end of cable poking out of the body of the sleeve, and above all to slide on and shrink with an industrial hot air gun a heat shrinking sheath, of the kind used by electricians. Those made by 3M, with resin sealant that expands with the heat, are the best.

Another practical note. Before pressing the sleeve, it is a good idea to place it in the jaws of the pincers and exert a slight pressure while at the same time pulling the cable until it forms the size of eye required, otherwise the tendency of the cable to slip out of the sleeve when it is pulled will drive you crazy! You can then finish the job by carrying out the necessary pressings.

Frequently the talurit are in aluminium and should just as frequently be studiously avoided for use at sea, for aluminium (the sleeve) plus steel (the cable) plus salt water add up to a boatload of corrosion! It is best to use those in copper or, in the case of nico-press, those in cadmium copper or nickel copper; the latter are by far the best. There is also nico-press in stainless steel, made by the American Loos company, a real authority on cables and terminals.

Once the diameter of the wire rope to use for the genoa sheet has been decided, and with it the type of terminal to fit (see j-lock)[18], we need to think about how to handle the wire rope. And that brings us to the problems of the polyester sheet and how to splice it.

[18] Today press-locks and Sparcraft j-locks are produced by an American rigging company, Aramid Rigging: another version is produced by *Tylaska*, also American.
[19] A line is tapered to make it easier to thread into a splice. Tapering also serves to avoid sudden changes in thickness where stress can be concentrated and lead to the splice breaking.

Common sense tells us not to handle a wire rope with our bare hands, and for this reason a much more 'user friendly' line is spliced to its extremity. Choosing the diameter of this line is quite simple: it should be of twice the diameter of the wire rope it is to be spliced to or one size up from that. To make this clear with an example, wire rope 5 mm in diameter requires a line with a minimum diameter of 10 mm or, better still, 12 mm.

The reason it is better to opt for a 12 mm line rather than a 10 mm line is a practical one: the wire-to-rope splice involves first splicing the core of the double braid line to the wire rope and then splicing the outer sleeve of the line to the wire rope. This involves passing the outer sleeve over the splice joining the core to the wire rope, so a slightly greater diameter makes this job, which is far from easy, less difficult.

Another tip on this subject is to cut away the core of the wire rope before tapering it. This avoids the not inconsiderable risk of the core, which is usually fairly rigid, piercing the outer sleeve of the line, especially when the line is rove through a block.

The wire-to-rope splice is a kind of splice that can be explained by patient souls with beautiful photographs and impeccable drawings, but is absolutely impossible to learn from photographs, drawings and explanations of this kind!

The skills needed in the many passages and the small but decisive tricks of the trade, born of experience and honed by constant practice, can only be acquired through almost infinite repetitions of the splice.

Once the hurdle of the wire-to-rope splice has been overcome, we need to pay attention to the lengths[20] of the wire rope and the fabric line to use. If we use enough wire rope to reach from the genoa clew to just short of the winch, so as to minimise stretch when the sheet is under load, this sheet will be of no use with the numbers two, three and four genoas, which have a shorter foot than the number one and thus will put load not just on the wire rope but also and above all on a good length of the polyester line.[21]

Before exotic fibre lines became available, this problem was solved by preparing a pair of sheets for the number one genoa with little wire rope, and a

[20] Usually a genoa sheet is one and a half times as long as the boat. For example, a 9-m boat will have a genoa sheet of 13.5 m.

[21] Several wise riggers – worth mentioning – used to splice a 'tail' to the line in polyester. This was simply a line smaller in diameter than the principal one, of about 6 or 8 mm, which served just to haul in the first few metres of the sheet when there was practically no load, so the resistance of the principal line was not needed. It also ran more freely and was lighter and more economical. The downside was that a sheet with a tail of this kind could not be inverted once it started to wear out. Personally I have used all kinds of materials, even strange ones, for these tails: from simple water skiing tow rope, which is a very light and economical single braid in polypropylene, to the expensive but excellent Yalelight from Yale: this is a mix of spectra and olefinic fibre.

pair for the numbers three and four with enough wire rope to reach as far as the winch.

Note that in this latter case the wire rope was frequently hidden inside the external sleeve of the line to avoid the steel wire coming into contact with the mast and shrouds. The need for this started to disappear at the very beginning of the 1980s when a new line – Cup – made by the German company Gleistein became available. This was double braid polyester with a core made up of parallel threads.

This construction produced much less stretch than other double braid lines, since parallel threads do not undergo the compaction that woven threads inevitably do and which produces considerable stretch.

The price to pay for this improvement in stretch performance, apart from a small increase in cost, was that, especially over the long term, this kind of line hardened considerably and thus tended to form kinks, which is far from ideal in a genoa sheet.

But innovative technologies were on their way, and with the coming of the first line in exotic fibre – Kevlar – the most up-to-date riggers used this in place of polyester, and in the space of a couple of years entire sheets in *Kevlar* made their appearance, with not a scrap of wire rope in them, even for the pendant.[22]

We can officially date this great leap forward in running rigging with the victory of *Passion* in the 1984 One Ton Cup held in Trinité sur Mer: this boat designed by Philippe Briand had *all* its running rigging entirely in Kevlar, including the runner tails, while the remainder of the fleet, though highly competitive, was still using a mix of fabric line and wire rope.

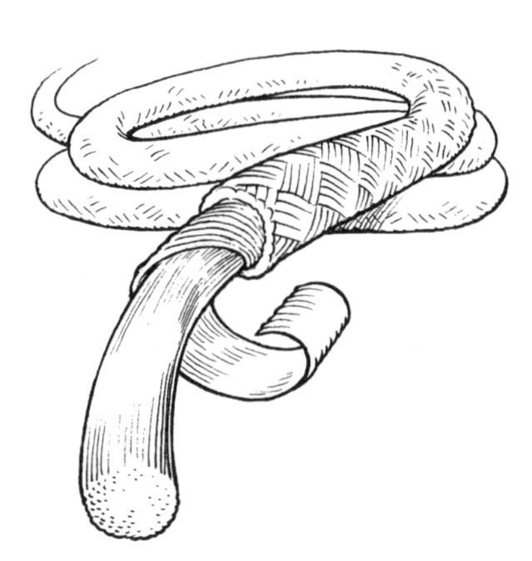

2.10
Gleisten Cup line.

[22] The very final part of the line, when it is in wire rope, is called the pendant.

With the advent of Kevlar, miles of galvanised steel wire rope ended up on the scrap heap – but the passage from the old to the new world cost three times as much as before! A wire and fibre rope genoa sheet cost a third of one in Kevlar when the new fibre made its debut! And today, with the use of Kevlar almost abandoned on board, that cost ratio remains if we compare the price of a genoa sheet in Spectra, not to mention Vectran or, even worse, PBO, to that of one in wire and fibre rope.

However, what Kevlar at the time (and nowadays all the exotic fibres that have followed it) had going for it was 'user-friendliness'. It was 'on the side' of the trimmer who had to use the sheet, making his job easier thanks to its superb handling qualities that were far above those of any sheet in wire and fibre rope.

And in addition, a sheet entirely in fibre is lighter than one using wire rope, and this made them very welcome on board racing boats.

But that is not all. Try dimensioning the sheets of a maxi yacht, let us say a 70 foot cruiser, using polyester lines. If you set out to use polyester to cut costs you will discover, with no little surprise, that you would actually spend less using sheets in Spectra, because polyester line would have obliged you to opt for huge cars for the genoa sheets because polyester line of very large diameter demands very big sheaves!

By using lines in exotic fibres we quickly freed ourselves of the last remnants of wire rope on board, *et voilà* we had running rigging all in fibre. But then the pain started! The classic splice used on double braid lines was the Samson, named after the glorious American company of the same name that brought in the age of double braid line back in the 1950s,[23] gradually replacing the older twisted rope in all applications.

This splice, which was perfectly in line with American Army regulations, with instructions describing in great detail the quantity and position of threads to cut away when tapering the line, was born for use on polyester line where both the core and the external sleeve (the actual braid) were designed to bear 50% of the working load. On the other hand, exotic line came into being with the avowed intention to have 90% of the working load borne by the core and only the remaining 10% by the sleeve, which is usually in polyester.

Thus, using a Samson type splice for exotic fibres was a necessary and inevitable straining of the rules, but certainly was not that appropriate. The first to realise it were, as usual, the America's Cup riggers, and especially Peter Gardner, an Australian who in 1983 had rigged *Australia II*, for the Newport races.

[23] American aircraft carriers moored in the Gulf of Taranto back at the beginning of the 1950s with double braid mooring lines. The space shuttle *Gemini*, when it was recovered from the ocean, was not hoisted on board with metal ropes but with double braid Samson lines.

Gardner, aiming to splice Kevlar lines exploiting the full breaking strain of the strong Kevlar core, developed a kind of splice known as the 'Australian splice', which involved splicing about a metre of the two cores together with the same simple method adopted for the single braid polypropylene lines used for water skiing tow lines or lifebelts!

The apparent complication of having to work on a double braid line was got around simply by cutting the external polyester sleeve with a soldering iron, right where the eye was to be spliced: at this point the tapered core was spliced back into itself for a metre and the external sleeve put back where it had been cut (this did not damage the Kevlar core since it was fireproof) and the join patched up by stitching and applying a leather sleeve. Later the leather was replaced by a heat shrink sheath.

This kind of splice pulled everyone out of the tangles of the wire-to-rope splice with all its difficulties and opened the doors for legions of makeshift riggers who laid claim to the name on the sole grounds that they were able to do an easy-peasy Australian splice. Those who embraced this (pseudo) innovation, splicing only Kevlar – and only with that method[24] – may have swept away the traditional splices but they were building their house on sand, forgetting that there is no progress without respect for tradition!

Another kind of splice was soon to emerge: the Gillespie, which was difficult enough to sort out competent riggers from the rest of the pack. But let us go back to Kevlar for a moment: if on the one hand the Australian splice had the undisputed advantage of being a recipe for 'cooking' the new exotic material without going back to the old methods used on polyester line, on the other hand it was not without its defects. A metre of Kevlar core spliced into itself transformed even the supplest line into a hard, stiff sausage, especially around the splice itself.

And let us not mention the fact that the leather or synthetic sheath over the splice only lasted a moment and continually required repair or replacement.

Terry Gillespie, of the New Zealand rigging company that bears his name, made the Gillespie splice popular. This was based on the same principle as the Australian – the exotic core was spliced into itself – but in addition exploited the outer polyester sleeve, recuperating it with a tricky manoeuvre using a special marlinspike. Since the Gillespie uses 50 cm of core and there is no need for a

[24] Other riggers, myself included, even with Kevlar lines with an external sleeve that was very finely woven and thus very hard to work with, continued to use the Samson splice so as to have more flexibility in the part where the line was spliced: the Achilles' heel of this option came out on the all-fibre spinnaker afterguys on maxi yachts, where the Samson splice showed all its failings. A long time afterwards, when I had the honour of welcoming to the marine cordage workshop I ran with Vittorio Vongher, Raymond of Spencer Rigging, who had come to Italy to work on *Barbarossa*, I discovered that other international riggers had had this same problem, and it was solved by an exchange of valuable information.

covering since the sleeve goes back into the line as in the old Samson, the result is a much more flexible splice and one that also looks a lot better, which does no harm.

On this point we should remember that the excessive rigidity of the Australian splice made the join vulnerable in the area where the splice of the core into itself finished and the line in its original, single core form began again. This was the point where all Kevlar lines, and especially the genoa sheets, broke. Decidedly exasperated by continual breakages on board the first two *Brava* One Tonners, I was forced to start taking note of the working hours of the genoa sheets and 'retire' those that had reached a certain working life. The fact was that Kevlar, although it had excellent mechanical characteristics in terms of breaking strain and stretch, suffered terribly from both exposure to solar radiation[25] and bending, even around a sheave.[26]

What is more, as if that were not enough, Kevlar fibres were also subject to abrasion. It is easy to verify this by sliding the Kevlar core out of the outer sleeve of a line that has been in use for some time and noting the cloud of yellowish powder that comes out with it: this is produced by the fibres rubbing against each other. Note too that the strands are opaque and no longer of the shiny yellow colour characteristic of new, unused fibre, and are covered in a kind of superficial fuzz made up of microfibres that have already broken. For all these reasons, Kevlar did not enjoy a long life for use in lines: at the 1985 Admiral's Cup there were already the first boats rigged with Spectra from Samson. The reason for the rapid success of Spectra, despite its high cost when it was first launched, lay in the weight saving it offered. In fact Spectra, being a polyethylene, has a specific gravity of less than one, which means it actually floats! Not only is Spectra lighter than Kevlar when dry, it also has the colossal advantage, since lines tend to get wet on board, of not absorbing water!

In truth, here we have spoken of Spectra and Kevlar as fibre, while in the real world the lines almost always have a sleeve (they certainly all did when these materials were first introduced), and this covering is usually in polyester. This somewhat levels out the difference, since the sleeves of both types of line will soak up water. But in any case, the weight advantage of Spectra remains.

Another basic reason for preferring Spectra is that it is not harmed by exposure to solar radiation, and this is no small advantage!

At the 1986 One Ton Cup in Palma di Maiorca, the German fleet arrived with lines in Spectra, made by the German company Liros, that had no sleeve. The

[25] In my marine cordage workshop I used to cover the reels of Kevlar with black cloth, because simple exposure to the light aged the fibre. If the reels were not covered, after just a year the line hardened without ever having been used.

[26] Kevlar absolutely has to have flat section sheaves.

lines in question, the halyards, were made all along their length, from the winch to the head of the sail, in naked Spectra. Without the polyester sleeve that normally covered the core!

These halyards, although they lost a very small and insignificant part of their breaking strain without the polyester sleeve, acquired the enormous advantage of a smaller exterior diameter, and this made them run very freely over the sheaves. When you consider that reduced safety coefficients on the masts of racing boats mean very narrow swallows for the halyard sheaves, being able to use line of smaller diameter was considered manna from heaven.

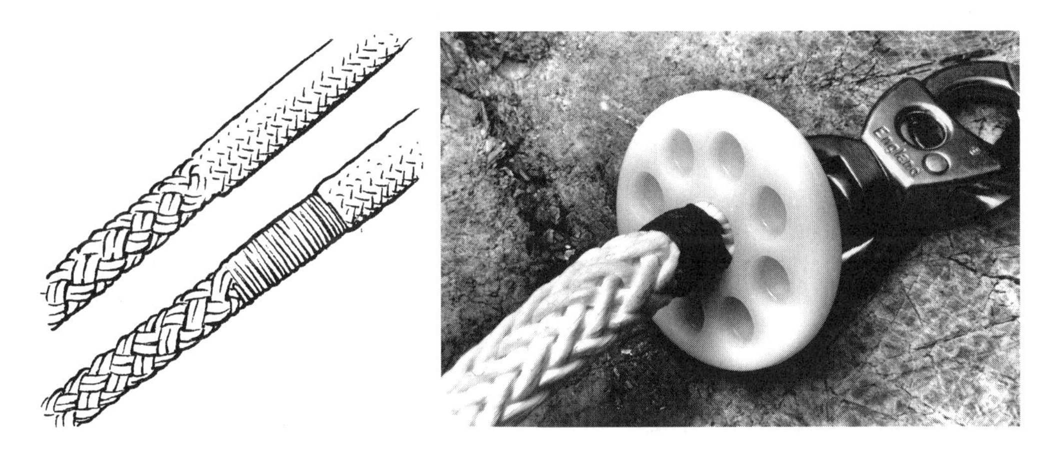

2.11

Lines for halyards with their sleeves removed; right, a single braid line in Spectra for a Spinnaker afterguy.

As if these first two major advantages were not enough, there were two more to add: Spectra[27] is in no way damaged by bending and is also easy to splice in its single braid form, using the water ski tow line technique.

And the disadvantages, you will ask? Well, certainly the usual rule applies: in nature you gain on one hand and lose on the other. In the case of Spectra, apart

[27] Some authors have spoken of Spectra Light when referring to Spectra, confusing what was merely the commercial name of a line produced by a cordage company with the material itself. This is like calling any low-calorie soft drink Coca-Cola Light™ In fact, taxonomic mistakes and suggestions of this kind are not uncommon, thanks to an unparalleled mix of ignorance and bad faith. When Vectran became popular some used to sell a line with the trade name of Vectra (made from Kevlar) as the more noble Vectran, while Spectra Light was merely a double braid line with a low content of spectra in its core, see Marlow spectra cruising. Also when Kevlar lines first came out, there were a lot of double braid lines with a few – just a few! – threads of Kevlar in the core. Obviously these were sold at almost the same price as proper Kevlar line.

from an initial cost decidedly higher than that of Kevlar, the biggest drawback was the so-called 'creep', which led to progressive stretching when under load: unlike Kevlar, which showed significant stretch at around 30% of breaking strain but stretched very little more as the load increased, with Spectra the amount of stretch continued to increase noticeably as the load got higher.

That is not all: for while Kevlar showed this behaviour right from the very first time it was used, Spectra – according to Samson's recommendations – needed 50 cycles of 'running in'. We can only guess what 50 cycles in a laboratory mean in terms of work on board! Attempts were made to get around the problem simply by pre-tensioning the line at 30% of its breaking strain before using it on board.

Because of Spectra's tendency to 'creep', Kevlar is still around, and has been for a long time, in racing sails, while Spectra has been confined to sails that are efficient but only for cruising boats, with one notable exception, a racing drifter[28] in Spectra.

To get a better understanding of the concept of creep, we will illustrate in more detail the concept of stretching in a line. A brand new line, put under load for the first time, will undergo what is called permanent stretching, caused by the orientation of the fibres and the weave of the braid, in the direction of the load applied.

It was for just this reason that the old Cup line made by Gleistein[29] that we saw earlier was so valid in its time, for the threads in the core were laid out parallel to the longitudinal axis of the line and thus underwent none of the stretch caused by the weave in normal braided line. The only permanent stretch came from the outer sleeve and the small amount of intermediate sleeve.

This permanent stretch is non-elastic lengthening, in the sense that once it happens there is no going back, it does not go away even if the load is taken off the line: basically, the line has become definitively longer than its original length.

We must also take account of the second kind of stretch, known as elastic stretch, that occurs every time the line is placed under load and immediately disappears every time the load is taken off. Thus with a new line – the classic case is a halyard fitted the evening before the race or before setting sail – we must consider total stretch as the sum of permanent stretch and elastic stretch, while with a used line we need only take elastic stretch into account.

[28] Drifter: a very light foresail with an area equal to or greater than that of the #1 genoa, designed to get the boat moving in calm or near calm conditions.
[29] The concept of this line has frequently been copied: see Sta-set from New England Ropes.

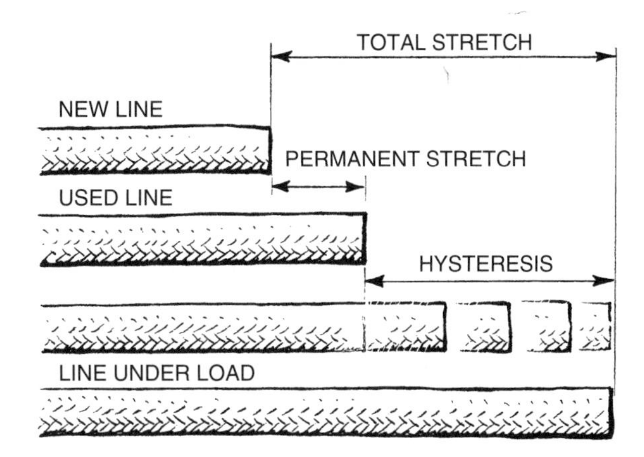

2.12
The various kinds of stretch on a line.

In the diagram on this page, note also the phenomenon of hysteresis, the portion of stretch in a line subjected to loads that take it outside its elastic range – that takes hours or even days to go away.

But we were talking about creep in Spectra; the fact is that the internal core of a line[30] in Spectra is very different in its mechanical behaviour from the external sleeve in polyester.

In this context, let us compare a three-strand rope with a double braid (noting that the three-strand has a smaller contact footprint on the sheave compared with the double braid). Thus it has a load per square millimetre higher than that of the double braid, and this makes for lower abrasion resistance load for load. This is why some kinds of Spectra, like Superbraid from Southern Ocean Ropes, which tend to flatten under load, are particularly long-lived and resistant over time. The same reasoning leads us to use flat section sheaves and flat sectioned lifting strops with Kevlar.

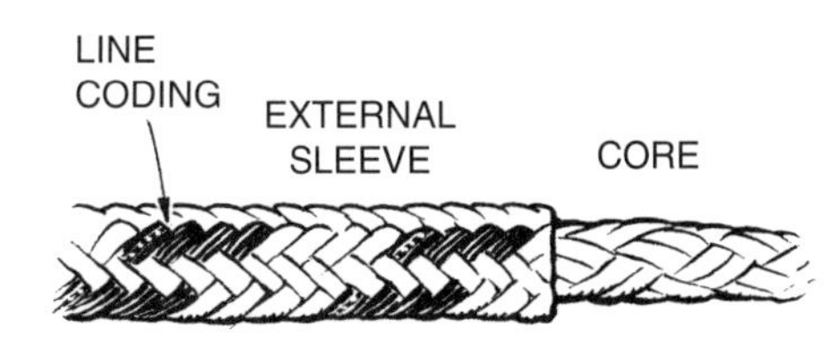

2.13
How a double braid Samson line is made.

[30] Modern double braid lines are made up of two pieces of line: one is the core and the other the outer sleeve. Sometimes between these two is a third element consisting of very light wound thread that may serve to contain a core with parallel fibres, as in Gleistein's Cup, or to prevent slip between the two braids in some brands of Spectra. Double braid line has a higher working load than traditional single braid with three, four or eight strands of the same diameter and material. In addition, it stretches much less because the strands go from one end of the line to the other at a less sharp angle.

The polyester sleeve, though it is only 'responsible' for 10% of the overall breaking strain of the line, with the remainder handled by the Spectra core, still perceives – and supports – the same working load as the Spectra core, but since it is inferior to the core in both quantitative and qualitative terms, cannot help stretching more, and often this stretching goes outside the elastic range and thus is still present when the load is taken off.

When the load is taken off, the Spectra core returns to its original length, while the polyester sleeve is decidedly longer than before because it has stretched to a level that cannot be recuperated. And since one of the braids is longer than the other, the sleeve is loose along the full length of the line.

Thus, while the Spectra core is working within its elastic range, meaning the stretch disappears once the load is taken off, the polyester sleeve is well outside its elastic range and so the stretch becomes permanent and does not disappear when the load is taken off.

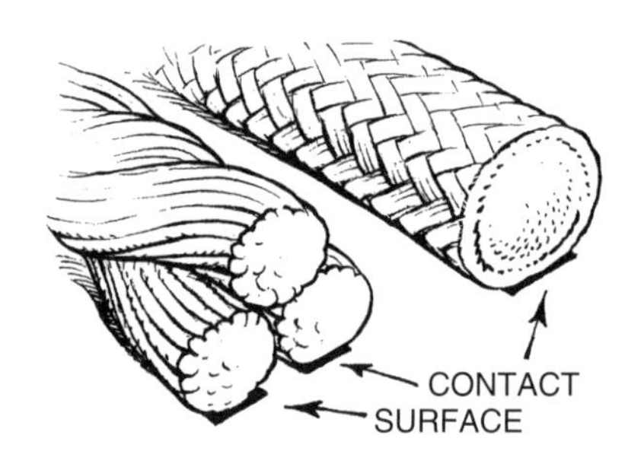

2.14

Three strand line. Right, double braided line: note the difference in the surfaces that come into contact with the sheave.

This causes a big problem that is wrongly referred to as 'slipping between braids in a Spectra line', and manifests itself with a polyester sleeve that is loose, swollen and baggy around the Spectra core. This leads to the sleeve breaking, quickly developing tears the first time it rubs against something, for example the jaws of a stopper. If we add to all this the phenomenon of 'creep' that manifests itself when the Spectra is subjected to high working loads for long periods or to sudden repeated high loads, we will end up with serious problems and a very loose outer sleeve.

The remedy? There is one but it is very tedious: we need literally to 'milk' the line from one end to the other, trying to push all the looseness in the sleeve to the end of the line and then 'purge' it; with the end of the line open, we literally cut it, and then finish off the end of the line again when the operation is over.

Together with Vittorio Vongher, in 1988, I solved this problem at source by adopting a simple but effective technique that we called 'floating', because the end of the line was left 'floating' and the looseness of the sleeve that is produced in Spectra lines, for the reasons we have seen, was automatically 'purged'.

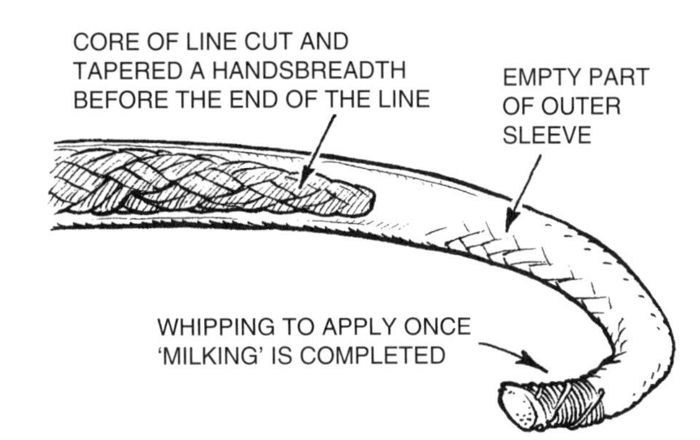

2.15
A method of finishing line with a 'floating' taper.

Other solutions have been tried to prevent the formation of looseness on the outer sleeve: for example, in an extremely radical approach to the concept of lines with exotic cores, attempts have been made to use the naked Spectra core as a line (we will look at an attempt with the spinnaker afterguys of a 64 footer in the chapter on spinnaker lines), since already back in the 1980s Samson was producing a line called Spectron, which was a single, 12-thread braid in Spectra that was compacted by chemical waxing[31] that held the threads firmly together. Here we will limit ourselves to saying that there was no way of making Spectron grip in a stopper or a jam cleat, and even less chance on the drum of a winch!

Just as we had to tackle the problem of runner tails being seared on the winch, now there is a similar problem with the genoa sheets on the winches of America's Cup boats, and this time the solution was found by an Italian company, Gottifredi & Maffioli, which made a special line with an exotic core and a sleeve in PBO. The long life of genoa sheets made from this line was recognised even by competitors of the Italian maker.

It has to be said, however, that on a cruising boat, where the keywords are simplicity and a certain attention to the wallet, with the 1142 kg calculated as the working load on a number three genoa in 40 knots of wind, or rather, on a roller jib reefed to the size of a number three genoa, a good polyester line fixed with a bowline to the clew of the genoa does its job very well.

Double braid polyester, of course, but of what diameter? Remember the result of the Marshall formula? The maximum working load on the genoa cringle was 1142 kg: the ideal would be 16 mm double braid polyester Yale Braid by Yale. In fact a working load of 1142 kg is just under 20% of its breaking strain of 6349 kg, so the line would be working in ideal conditions as defined by its maker.

[31] Yale was a forerunner with its Maxi Jacket Coating liquid for waxing line cores.

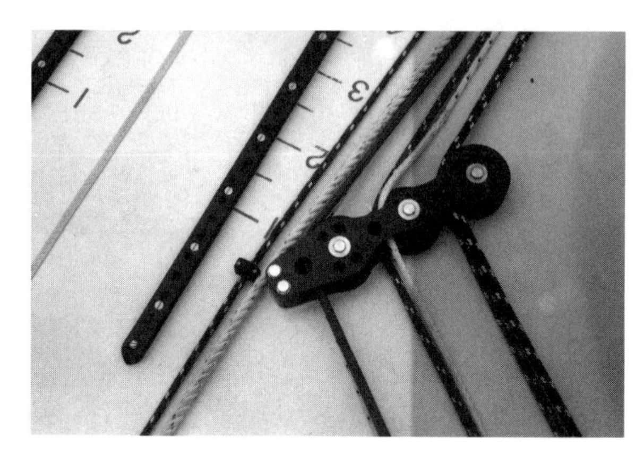

2.16
Southern Ocean Ropes lines.

At the same time, although the bowline reduces the breaking strain by 40 or 50%,[32] with 3174 kg[33] we still have a safety factor of three to one, which is a wise choice for a cruising boat. Want to know the stretch? The Samson table reproduced on this page helps us to find this. Along the *x*-axis are working loads of 10, 20 and 30% of the breaking strain, while on the *y*-axis are values for elongation of up to 10%. To the right of the table are shown the names of the various types of line tested, by being subjected 50 times to 10, 20 and 30% of their breaking strain, except for Dacron Yacht Braid. This line, represented by the topmost curve in the graph, in part because it has not been pre-stretched, and thus shows the permanent stretching we spoke of earlier, and in part because it is a 'totally anonymous' line and certainly not a top brand, already at 20% of its breaking strain shows a stretch of 8%!

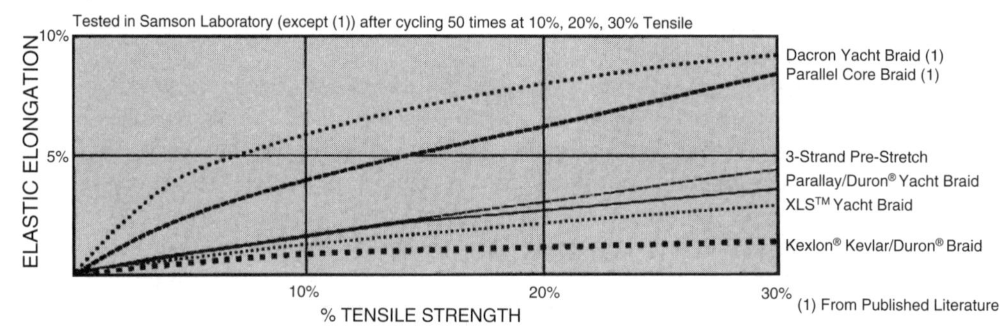

2.17
Graph from Samson for calculating line stretch.

[32] As with any other knot tied in a line.
[33] Half of 6349 kg.

On an IMX 38, with a distance between the clew of the number three genoa and the winch of about 3 m, a stretch of 8% would translate into 240 mm: like having an elastic band instead of a genoa sheet! We could ignore this if the IMX 38 were a cruising boat, but it would be unacceptable on a racing yacht.

The lowest curve in the graph shows the behaviour of Kexlon Kevlar, a double braid with an exotic core in Kevlar from Samson, and the stretch here at 20% of the breaking strain is down to a very low 2%, and over a length of 3 m under load, this would mean a stretch of just 60 mm!

The genoa cars

How to Dimension Them

Cars Adjustable Under Load and Those not Adjustable Under Load

Track Systems: Fore and Aft and Athwartships: Manually or Hydraulically Operated

When we were talking about genoa sheets in Kevlar and about their sudden and irritating breakages, we mentioned how much Kevlar hated bends. Well, in the number three genoa car, the sheet is turned through an angle of as much as 70–75°, so it was certainly no accident if all the sheets gave way at this point where the car turned them through the greatest angle. If we put this fact – a turn through 70° means a load increase factor of 1.22 – together with the fact that the number three genoa produces the highest load of all, it is easy to understand that the number three genoa car must be dimensioned with special care!

Well – to continue with the example of the IMX 38 – where we dimensioned the genoa sheet on the basis of the 1142 kg that the Marshall formula produced, we now have to multiply this by 1.22, because of the angle by which the sheet is turned with the number three genoa. And this gives us a working load on the car of 1393 kg.

To underline how important it is to take into account the angle of deflection, let us work again through our example, still with a working load of 1142 kg but supposing that this is the load calculated for the number one genoa clew. In this case the angle of deflection of the number one genoa sheet would be 45°, so the load factor would be 0.76 and the load on the car would actually be less (1142 × 0.76 = 867) than that on the clew.

What should we do in this case? I mean, if the loads are different, and they certainly are because not only does the number one genoa car bear a lower load

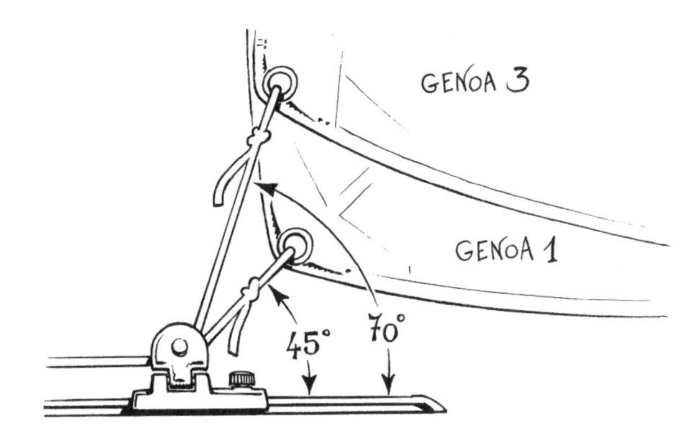

3.1
Angles at which genoa sheets work (note the 70° for the genoa #3).

than the number three because the angle of deflection reduces the load present at the clew, but also this load is less in the first place. The load on the clew of the number one is less than that on the clew of the number three. So which car should we use? Two of the same kind designed to bear the loads of the number three genoa but well over-dimensioned for the number one genoa, or a stronger car for the number three and a less strong one for the number one?

This question does not have a single answer, but directly implies another question: cars that can be adjusted under load or not? To tell the truth, the apparent enigma is soon resolved, for adjustable cars cost a great deal more, four times as much as a non-adjustable system, so the question answers itself. For a cruising boat, where we must be guided by a reduced budget, the best choice is to use non-adjustable cars, while on a racing boat it is obligatory to have cars that are adjustable under load, not so much because money is no object but because it is essential for optimising boat speed to be able to adjust the cars constantly and continually so as to give the genoa an optimal shape.

On a cruising boat the preference is for the extreme simplicity of a single track that accommodates all positions for the sheets of the various genoas (or the various positions required for a single roller reefed genoa) with the cars dimensioned for the number three genoa. On the racing boat, the tendency is to have cars with a working load less than that of the number three genoa for numbers one and two, not for cost reasons but just to save weight!

Here too, needless to say, there are exceptions. First of all, on a cruising boat that is not conditioned to extremes by cost considerations, it is perfectly right to consider the option of using a system of genoa cars that can be adjusted under load. This is because when the area of the genoa is reduced or increased, either by a sail change or, much more commonly, by a roller reefing system, the car that can be adjusted under load, and thus manoeuvred from the cockpit, completes the sense of safety that roller reefing provides, since we are not obliged to leave the shelter of the cockpit.[1]

On a racing boat, where these difficulties do not exist because the crew is larger, often the choice is for number three genoa cars on simple 'T'[2] tracks, as these are slightly lighter than adjustable ones mounted on 'X' tracks.

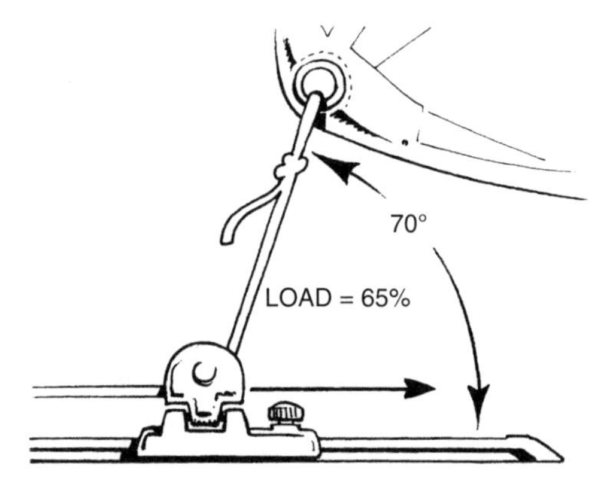

3.2
To move the car – when it turns the genoa sheet – you need to apply a load equal to 65% of that on the genoa sheet.

It must also be taken into account that the number three genoa is less used than the number one, and that the adjustment of a number three genoa car is a hard task because of the high loads involved. But what exactly are these loads? In the case of the number three genoa, to adjust the car under load we have to apply a force equal to 65% of that on the sheet.

In the case of the IMX 38 we would need to apply a force[3] of 742 kg to move the car when the number three is flown in 40 knots of apparent wind. We would need a 20 : 1 purchase and still have to apply a force of 37 kg, which is no small effort (even without taking into consideration the loss of efficiency through friction in the purchase . . .)!

[1] A good wrinkle for getting round the problem of adjusting the genoa car once you have changed the area of the foresail, in the case of genoa cars that cannot be adjusted under load, is to fit the cars with a control line leading into the cockpit and use this to slide the weather-side genoa car, where the sheet is not under load, into the desired position, then use the roller reef mechanism to change sail area and go about, thus ending up on the new tack with the car in the right position. From this you can well understand the importance of deciding carefully while the boat is being built on what kind of genoa cars to fit, because changing afterwards from non-adjustable 'T' tracks to adjustable 'X' tracks will not be either easy or cheap.
[2] 'T' tracks are the traditional ones with cars that run on plastic guides. These guides are necessary because without them the contact between the metal of the car and that of the track would wear away the anodising or anyway eat into the metals; the 'X' track is the classic one for cars on ball or roller bearings.
[3] 742 kg is 65% of 1142 kg.

Fortunately, in reality things are a little better, because in a puff of 40 knots nobody would dream of moving the car forward, if anything it would be moved aft to open up the leech and depower the genoa! The car would be moved forward in the lulls between one gust and the next.

Attention also needs to be paid to the control block that leads the control line aft into the cockpit. This turns the line through 180° and thus doubles the working load present on it. In our case we would need a block with a minimum working load of 1,484 kg.

In what way do cars adjustable under load differ from non-adjustable ones? The former have ball bearings circulating freely in a 'race'; the large number of balls (several dozen for the car of a 40 footer) allows the necessary working load[4] to be achieved, while at the same time the shape of the balls means they have a very small contact footprint both on the car and on the track, and this means reduced friction. It is this reduced friction that allows the car to be moved under load. Traditional cars, not adjustable under load, run on a 'T'-shaped track with simple guides in plastic material to reduce friction between car and track. But to tell the truth, there are compromise solutions in both types.

For cars not adjustable under load, recently several producers have tried to improve the quality of the guides by using special self-lubricating plastics: the smoothness of the cars is better than in traditional models but still less than that of cars with bearings, and unfortunately the cost is very close to that of bearing systems, so it is a compromise.

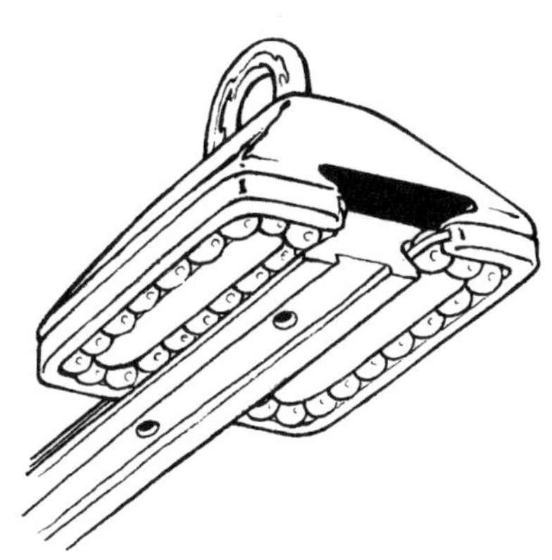

3.3
A series of ball bearings in a race inside a car.

In cars adjustable under load, when ball bearings are no longer sufficient to handle the working load, the preference is to use in their place cylinders, also called 'rods', of Torlon[5] which, while they do not provide the smoothness of ball bearings under low loads, allow the car to be adjusted under high loads without undergoing the deformation balls are subject to under extreme pressure. After the 2000

[4] The greater the number of ball bearings, the greater the working load.
[5] Torlon is a plastic material that needs no lubrication; it is also called Duratron.

America's Cup, new cars became available that were made by Harken in collaboration with Jack Roeser of Otto Engineering. These were based on Torlon roller bearings that circulated in the usual race, but the peculiarity of these cars was not only that the roller bearings were captive, and thus did not slide out if the car was taken off the track, but also the fact that thanks to the roller bearings these cars reached incredible working loads. For example, if you see how small these cars look when mounted on a maxi yacht, you may think a mistake has been made.

3.4
Rondal genoa cars for cruisers with 'T' tracks.

As with everything, cars with roller bearings too have their pros and cons: certainly despite huge working and breaking loads they do not provide the smoothness at low loads that is still the privilege of cars with ball bearings. They are also particularly susceptible to accidental bending of the track and take this decidedly badly. So it is really important that the track is mounted flat on the deck, something that is often taken for granted but in practice is not always the case! The alternative is to use not a single car but a 'train' of smaller cars, as on the French 60 foot trimarans.

Ball bearings in cars may also be in Delrin, a plastic material with less mechanical resistance than Torlon but cheaper, though it also tends to wear out in the long term under the action of atmospheric agents.

Usually ball bearings in cars do not require specific maintenance other than a rinse in fresh water after every outing and about once a year a good clean to get rid of dirt and salt crusting by removing the car using the loader[6] and submerging

[6] The loader is a very useful accessory to keep always on board. It is simply a piece of track-shaped plastic that lets you remove the car without losing the ball bearings, by placing it at the end of the track after removing the stoppers, and sliding the car on to it. Once the car is on the loader, be sure to lock it in place with the spring clips. The loader will also be useful for reinserting any ball bearings that fall out; it has a special opening for this purpose.

it in a bucket of water with some washing-up liquid, cleaning the balls and their race with the help of a brush.

Types of car? These range from the typical car with ball bearings for racing boats to the common car that may have either a laterally hinged sheave or a fixed sheave forming part of the body of the car, as produced both by Lewmar and Harken in the models known as 'tri-roller'.[7]

3.5
Barbarossa 'tri-roller' genoa car for 'T' tracks not adjustable under load.

The cars may be controlled by a purchase, a winch or a hydraulic system. For example, the Harken adjustable car with ball bearings, for boats of around 24 feet, runs on a fore-and-aft track. The car is pulled astern by a strong shock cord if the genoa sheet tends to pull it forward, and is adjusted in a forward direction by a 2 : 1 purchase with a control line leading to a jam cleat in the cockpit.

As the size of the boat increases, the purchase needs to be more powerful to be able to move the car. The extreme application of the purchase to move the car comes on boats between 45 and 50 feet, where sometimes instead of a purchase the control line is led directly to a winch. On larger boats, and especially on maxi yachts, a hydraulic cylinder provides the power to move the genoa car.

To fly the genoa in a more open position[8] compared with that imposed by the main track, a second track is installed parallel to but in a more outboard position than the first. A track external to the main one does allow us to open the genoa

[7] The *tri-roller* is an invention of the Italian company Barbarossa, since taken over by Harken. The tri-roller seems to have originated from an intuition of the Bassani brothers (one of them is currently owner of *Wally*, but at the time they were joint owners of Barbarossa, as well as keen sailors, and they commissioned their designer Luciano Bonassi to design a car that 'did not slam into the sides of the cabin'. Since then the tri-roller has been copied all over the world.
[8] By a 'more open' genoa we mean one which, with identical sheet tension, forms a wider channel between itself and the main.

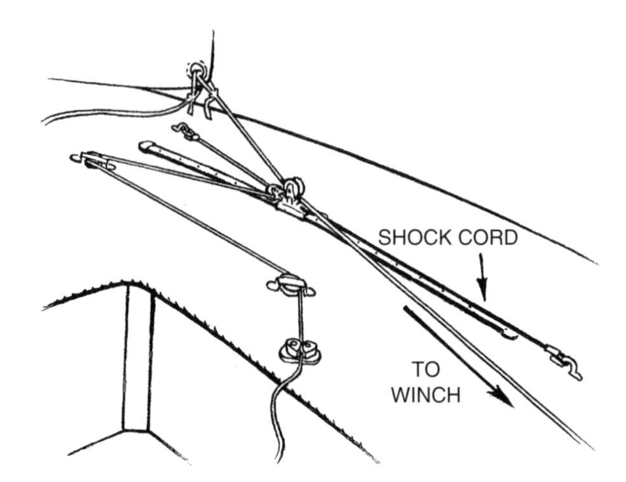

3.6
Harken car with bearings, adjustable under load with a 2 : 1 purchase, for 24 footers.

more, and this is useful if there is a sudden increase in wind strength that we do not think is going to last and thus does not merit a sail change (or if we have to sail a tack further off the wind but not enough to be able to fly the spinnaker or asymmetric chute, yet still want as much canvas as possible forward with a sheeting angle that is well open), but opening the genoa in this way is a laborious business.

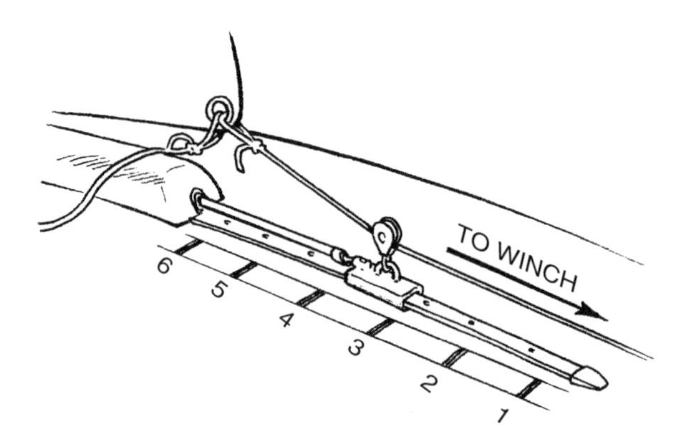

3.7
Maxi yacht genoa car adjusted by a hydraulic cylinder.

In fact, to open the genoa we have to bend on, or have permanently in position, a 'short sheet', a sheet much shorter than the usual one (since it will never be used during a tack) with an open Wichard shackle in place of the j-lock. This will be temporarily clipped on to the genoa clew so that the genoa can be opened by tensioning the short sheet that will be rove to a car or fed into a snatch block[9]

[9] Here we mean a snatch block, a block whose cheek can be opened, with a snap shackle on its base.

positioned outboard of where we want to have our genoa clew. This inconvenience can be eliminated by using a genoa track system with athwartships adjustment.

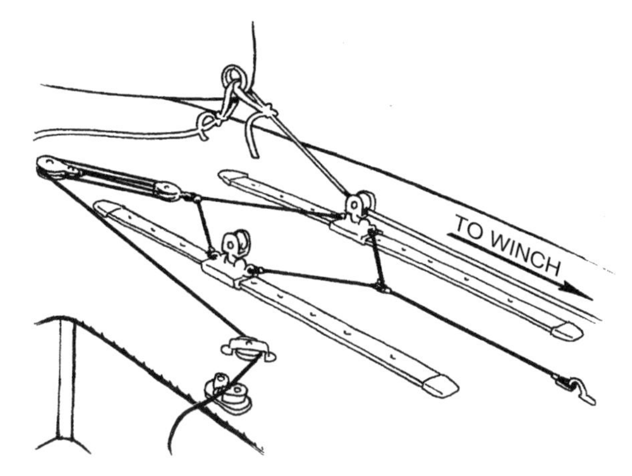

3.8
Ball bearing car system, adjustable under load with a 4 : 1 purchase, with external track.

In the 1976 Half Ton Cup,[10] the half-tonner *North Star* was the first to use athwartships tracks to regulate the genoa sheets.

3.9
Harken 'Custom' snatch block for maxi yachts.

Figure 3.10 shows an athwartships genoa car system for boats of up to 40 feet. The system consists of a track for the number one and two genoas and one

[10] Half Ton Cup was the world championship of the Half Tonners: IOR class five boats.

further forward for the three and four genoas. On each track is a ball bearing car that can be hauled inboard using a Harken Magic Box, while it moves outboard naturally under the pull of the genoa and also because the control block of the genoa sheet is purposely mounted in an outboard position so that there is no need for an additional control to pull the cars in an outboard direction.

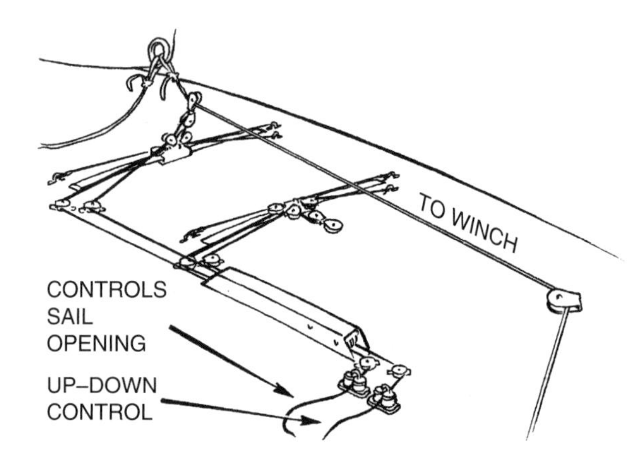

3.10

Tranversely adjustable genoa car system with magic box purchase.

This inboard–outboard control function is simply a much faster and, importantly, a gradual method of moving from an internal to an external track and vice versa. The second function of this system is the 'up and down' control, which allows us to regulate the height of the block through which the genoa sheet passes: the higher the block the further aft the sheeting position and vice versa. This control is indispensable because otherwise athwartships tracks would not allow the sheeting position to be adjusted in the fore-aft direction.[11] The advantage of the system is that it permits a virtually infinite range of adjustment; on the downside it is not at all easy to implement, and the genoa trimmer needs to be very skilled fully to exploit the potential of the system. A system designed for an 80 foot Bill Tripp cruiser racer, coordinated by Vittorio Mariani, had a special refinement: the track was inclined so that it would remain perpendicular to the pull on the block where the genoa sheet was rove so as not to work just on one side of the car, which is what happens when the pull is not perpendicular to the track. The inclination of the track meant that the end controls had to be realigned to make the control lines run parallel to the deck.

[11] This adjustment is important also if the rake of the mast is changed, for this too changes the sheeting point. You can see the rake in the sail plan: it is the amount by which the masthead is shifted aft with respect to a position at right-angles to the deck.

3.11
Transverse adjustment system for the #3 genoa of the VOR 60 footer *Illbruck Pinta* in the Volvo Ocean Race (note the shock cord holding up the flying block fixed to the diagonal shroud).

Then there was an interesting system of athwartships tracks used on a Vallicelli 44 footer, in which the up–down control of the genoa sheet was not designed to work in an athwartships but in the fore–aft sense. This solution simplified the various up–down controls and the attachments on the car and the end controls, but had the disadvantage that when the genoa car was moved inboard or outboard, the tension on the up–down control was also modified. In fact, since the track was straight, the car felt a tensioning every time it was moved off the centre line of the track.

The Lewmar system (Fig. 3.12) does have two perfectly fore-and-aft oriented tracks but is in reality a mixed system. The sheeting position is adjusted fore and aft, as in traditional tracks (though here with a hydraulic system), but the sheeting angle of the genoa is laterally adjusted by a hydraulic cylinder concealed in the tube connecting the two tracks on which a tri-roller type car is mounted, and this brings it fully into the athwartships adjustment type of system.

In recent years there has been a boom in self-tacking type genoa tracks.

The reason is quite simple: in pursuit of the 'easy sailing' philosophy, to get out of the complications that have for long hampered sailors, the Wally yard, headed by Luca Bassani, has pioneered this new way of sailing that attempts to reconcile the apparently contradictory requirements for enjoyment, speed, safety and user friendliness.

Self-tacking jibs are thus a not-to-be-neglected part of this concept that is currently being employed on fast cruising maxi yachts, with a significant impact on the products of the world's leading yards. Self-tacking systems made their debut back at the beginning of the 1970s on Elvstroem's *Solings*. The system consists of a curved track positioned athwartships and fixed to supports. The track brings the

3.12
Hydraulically operated integrated Lewmar system for transversely adjustable genoa car.

genoa car perpendicular to the pull on the sheet so that it can work correctly with all its rows of ball bearings.[12]

The reason for having a curved track lies in the fact that such a track behaves as the arc of a circle, as if traced by a radius. Since this radius is of constant dimension, the tension on the jib sheet, which is in fact the radius, will not vary as the position of the car on the track changes. The car thus runs freely on the track and assumes the ideal position according to the wind direction. Interestingly, the recently launched Wally 143 has, unusually and for structural and aesthetic reasons, a straight track for its self-tacking jib!

The jib sheet consists of a line coming from the bow (turned by a footblock or a block built in to the deck, in the case of a dinghy) rove through a block on the centre of the car and then bent to the clew of the sail.

The innovation brought in by Wally was to adopt the self-tacking system on big boats, from 70 to 100 feet, with a system that was decidedly more evolved than that of the self tacking foresail (with or without boom) seen on classic boats. In addition Wally also favoured the 100% genoa, with the great advantage of not having a genoa clew thrashing around amidships with every tack and risking hitting the guests on board. This advantage, together with that of self-tacking, obviously makes for a good system. The system used on the *Soling* had the not

[12] Ball bearing cars have a line of ball bearings on each side of the car that run along the sides of the track. All the ball bearings contribute equally to supporting the load on the car: if the pull on the car is not perpendicular, one line of ball bearings will bear a bigger load, and the more off-centre the pull is the greater this load will be so, at the very least, the ball bearings carrying the bigger load will wear out faster; and above all, the car will run less smoothly and working and breaking loads will be diminished.

3.13
Transverse system on a Vallicelli 44 footer.

inconsiderable disadvantage of having the jib sheet originating in the bow, so it swept along the foredeck, and this would be intolerable on an offshore boat.

Rather than using a lateral system of lines and blocks to control the genoa clew, the classic 'up and down' we have already seen on the athwartships tracks of maxi yachts, Wally devised a control for the genoa clew that came from above, simply by making use of a halyard. This halyard, which in fact is a control line, leaves the mast high up and, running through high load sheaves incorporated in the mast, comes down towards one of the two cars on the self-tacking track, is then rove through a block on the genoa clew, returns down to the track where it is rove through the other car and then passes back up through the mast.

This system has the advantage of leaving the foredeck unencumbered and also has another very important advantage: since the control line has two ends, one goes to a winch for fast hauling in and the other ends up on a hydraulic cylinder that is used for fine control. This double control system is also a back-up system in case of malfunction of either winch or hydraulics.

And its defects? Perhaps only one, or two. First of all, do not forget that an athwartships track of this kind cuts the deck in two forward of the mast, and it is not always desirable to have a car that can sweep away anything it finds in its path, people included! Another is that the cars run only too smoothly! When tacking, the cars shoot from one end of the track to the other with great speed and force, and when they arrive at the other end smash violently into the endstops, and this can lead to the breakage of the plate connecting the two cars.

Some may raise objections over the need for two cars, why not just have one? The fact is that a curved track,[13] and that of the self-tacking system is always well curved, would not allow a single car to run along it, because this car would have to be long to take the load (remember? The more ball bearings, the higher the load!). A long car just cannot run along a very curved track.

So this problem is got around by using two cars to bear the designed load, so there have to be two control lines going to the genoa clew. Initially, to hold the two cars together, the coupling always used for the cars on the mainsheet traveller was adopted, but this ignored the fact that on the main traveller it worked under traction, while on the self-tacking track it worked under compression and just was not up to it.

Actually, it has never been clear why a coupling should have been used on the self-tacking system, since the cars are in any case held together by the control lines coming from the genoa clew.[14] Subsequently the problem was solved simply by transforming the fixing holes of the coupling into oval openings so that when the car was brought to a halt the coupling did not put a strain on its fixing bolts and the shock was taken by the ends of the cars which had 'fenders' in plastic material.

Another 'do-it-yourself' solution that is just as effective is simply to join the two cars with a shock cord that keeps them together when needed but allows them to separate when under load.

[13] Curved track is ordered as a one off, by specifying the measures of the chord and its depth plus the type of track and curvature. You can in fact have track that is curved both in the horizontal and vertical planes.

[14] This is obvious so long as the genoa circuit is fore-aft, because otherwise, if it has an athwartships layout, the bar is obligatory.

3.14
Genoa cars with plenty of toggle.

Speaking of genoa sheet control systems, it should be pointed out that recently there has been an attempt to eliminate the flat block and place the winch in line with the track, or with the guide block of the transverse track or the self-tacking track. It is easy to see why: this flat block usually turns the genoa sheet through an angle of close to 120°, with an increase in the load of 72% (and in a point where the load is already high, see above on number three genoa sheets), and this means a very cumbersome and costly block is required.

Several yards that mass produce boats have opted for a simple car at the aft end of the genoa track so that the sheet can be led directly to the winch. This is a fairly effective solution and it is certainly economical, except for the fact that on the starboard side it is easier to implement since on that side the sheet passes

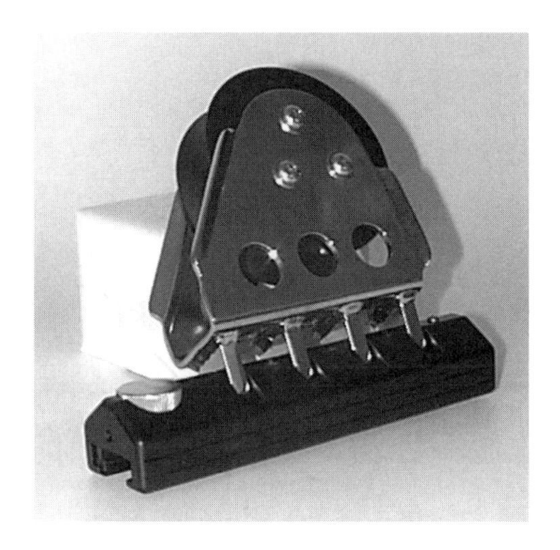

3.15
Toggled genoa car.

first round the outboard side of the winch drum. But on the port side, since winches are normally right-handed,[15] the genoa sheet has to pass first around the inboard side of the drum, and this change in angle can prove too much for the genoa car.

Shown in Fig. 3.14 is a solution proposed in the 1970s by the American Nicro Marine, using 'T' tracks with car and lead car both pivoting to lead the sheet to the winch.

[15] Winches usually rotate clockwise, as seen from the top. Exceptions are winches on America's Cup boats and some maxi yachts which are left-handed – that is, they rotate anti-clockwise – like those on certain cruising boats designed to create symmetry in deck layout or used because of lack of space in the walkways to the sides of the cabin, as in the case of Nauticat motor sailors.

The mainsheet

Tensioning systems: the eternal struggle of power against speed!

The block and tackle, which was already known to Leonardo da Vinci and studied by him, is still in vogue, despite the fact that certain writers seem to think it has been forgotten. From the smallest dinghies to maxi and mega yachts, it is always useful and irreplaceable. Using a purchase to tension a line instead of a system using a winch means having a system that is more economical, lighter, more flexible and sometimes even faster.[1]

Obviously there are limitations to the use of purchases, as the working load gradually increases to values that cannot be handled by a single purchase and need one or more winches as well, and there are also limitations to the use of a purchase as the sole and single means to tension a line.

Let us look at when and how to use a purchase system for the mainsheet.

Even though an unavoidable law of physics, the law that links the power[2] of a purchase to its speed, is a natural law that has to be respected: what you gain on the one hand you lose on the other. There is always a piece missing in life! If you construct a purchase that is particularly powerful, it will be inexorably slow, while one that is very fast will have little power. It is a question, as common sense will tell you, of finding the best combination, the right compromise between the two factors.

A 4:1 purchase, like the one shown in Fig. 4.1, with four parts at the moving blocks, allows 40 kg (let us say this is the working load on a boom) to be hauled with a force of just 10 kg, but at the same time forces the trimmer to haul 4 m of line to move the boom 1 m towards the amidships position.

But often such a simple purchase cannot meet the various needs that arise at different times on board. For example, a J24 needs a 4:1 purchase to trim the main in a fresh breeze, but the need to haul the boom in quickly when gybing demands a 2:1 purchase that is much faster than a 4:1. To meet these various requirements a cascade tackle needs to be designed.

[1] See inverted purchases.

[2] We have no doubt that the decidedly improper use, from the point of view of physics terminology, of the terms power, force, speed and so on will have more than one engineer or academic with his teeth on edge, but this book is written for the general public and takes a practical approach typical of sailing books, and such factors impose these 'dissonances'. Our apologies to the more sensitive souls.

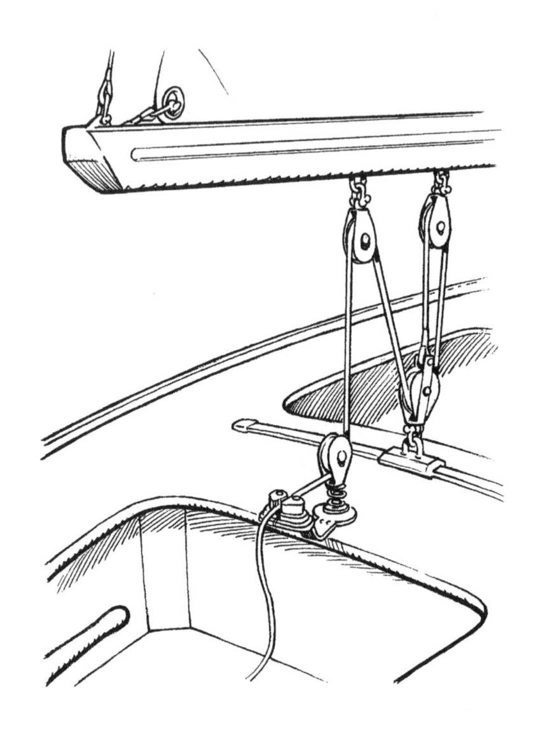

4.1
4 : 1 purchase with cam cleat.

Here, quite simply, a purchase is driven by another purchase. In a 'cascade' system the bigger purchase (note this is the one with the highest breaking strain) is used to haul in the sheet rapidly, while the smaller purchase (with a lower breaking strain, because it has to bear a load that is already divided by the number of parts of the main purchase) serves to provide the power to trim the sheet in a fresh breeze. The total mechanical advantage of a cascade system is equal to the number of parts in the main tackle multiplied by the number of parts in the secondary one.

For example, a 6 : 1 main tackle with a 4-part secondary tackle will give a purchase of 24 : 1. This means, in other words, that for every kilo of pull on the fine tuning part of the mainsheet, we exert a pull of 24 kg! Actually this is only true in theory, because it does not consider the friction that a block causes on every part. The more parts there are the more the friction increases, so it is a good idea not to get carried away and construct purchases with an enormous number of parts, because beyond a certain limit the paper advantage in terms of purchase loses its meaning as the friction increases tremendously.

To find out the number of parts in a block and tackle system, you need to count only the parts that come out of the moving block, in other words the number

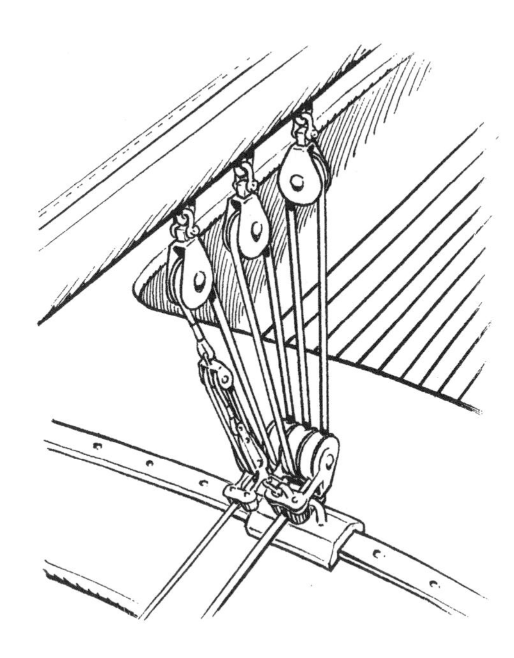

4.2
Cascade purchase with a total reduction of 24 : 1 with fast and fine tuning.

of parts below the boom or coming out of a car on the mainsheet track, or from the clew of a jib. Actually these figures need to be corrected for the friction present in the system, which means the performance of the system is never 100%. Let us take a detailed look, using as an example the circuit illustrated overleaf where there is a simple[3] 2 : 1 purchase, which thus has very few parts.

415 Kg
LOAD OF 395 Kg
425 Kg
405 Kg
436 Kg TO WINCH
800 Kg ON BLOCK

4.3
A mainsheet purchase with 2 : 1 reduction and the resulting increases in friction.

[3] This system has been the workhorse for the mainsheet of very many America's Cup international 12 m.

We can see that the 395 kg load on our hypothetical boat at the point where the sheet is fixed to the boom produces a load of 405 kg, once the sheet leaves the block on the mainsheet track because of friction, and each successive passage increases the load until it reaches 436 kg, a full 10% more than the original load of 395 kg. This means we must always bear in mind that every part added to a purchase will increase the load through friction.

If we want to calculate the length of sheet we haul in when trimming this 'cascade' purchase, we have to divide the length of the secondary purchase – the one used for fine tuning – by the number of parts in the main purchase used for fast trimming. The illustration here shows a classic fast-fine system for a 32 footer.

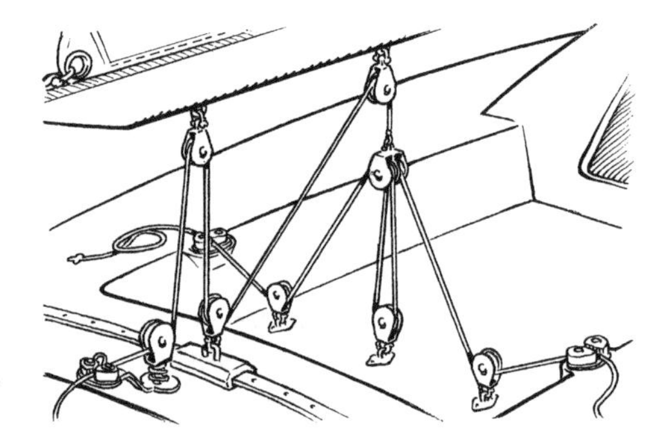

4.4

Mainsheet purchase with 'cascade' system for boats up to 32 feet. A 4 : 1 purchase is served by another, also 4 : 1, for a total reduction of 16 : 1.

A fine tuning 4 : 1 purchase acts on a fast 4 : 1 purchase, giving a final advantage of 16 : 1. In our case, the fine tuning purchase is 160 cm long and the fast purchase is a 4 : 1, so by using the fine tuning purchase we can shift the boom by 40 cm. These 40 cm may or not be enough: if they are not enough we will have to try to fit the fine tuning purchase inside the boom to be able to have an amount of regulation that allows us to move the boom by the necessary amount. We will have to 'try' because in reality this is easier said than done because of a series of problems, foremost among them that of our purchase tangling with the reef lines and the outhaul, if these are also housed inside the boom. A second point is that we need to evaluate carefully at what point to have the control line leaving the boom. It needs to be a point that the main trimmer can easily reach, even when the boat is heeled, and must not interfere with other pieces of rigging that are often in contact with the boom, for example the runners on a fractional rig.

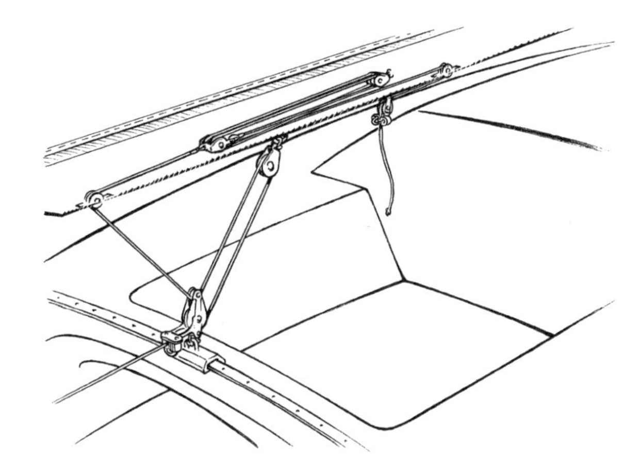

4.5
Mainsheet cascade purchase with fine tuning inside the boom.

Another limit to the fine-fast purchase system is reached when we stubbornly try to use it to trim a 'monster' main such as that of a One Tonner of the old IOR 30.5 class: from *Brava Les Copaines* (1985) to *Pinta* (1991). Although the main trimmer on *Pinta* was New Zealander Don Cowie, who is a hunk of a man, mainsails of 40–45 m^2 are unmanageable just with purchases. The purchase has of necessity to be used with one or more winches, and here we get back to what we said out the outset.

So far we have only looked at the possible power of purchases and not at how much power we actually require: how many parts does our purchase really need? If we look back at what we saw when talking about the purchase for the backstay of the Finot 34 footer, we had assumed a maximum hauling power for

Table 4.1 Recommended ratios for mainsheet purchases[4]

sail area (square metres)	purchase
9–15	2:1–4:1
15–20	4:1–8:1
20–25	4:1–16:1
25–30	4:1–16:1
30–35	6:1–24:1
40–45	6:1–36:1

[4] Marty Rieck, 'Speed versus power', pages 8 and 11, in *Technical sailing*, n. 2 Sept–Dec 1993.

a crewman[5] of 34 kg, and so with the 30:1 purchase we would need for a boat with a mainsail of 40–45 m^2, the theoretical pulling power would be 1020 kg, a value obtained by multiplying the crewman's 34 kg by the 30 parts in the purchase system we have. And this pull is 'theoretical' because it does not take friction into account. But note that we reach a working load of 1020 kg with a mainsail of 36 m^2 in 30 knots of apparent wind, so the figure in the table below would appear to be wrong, as a mainsail of 45 m^2 generates a maximum load of 1276 kg in 30 knots of wind. In fact it does not work like this, because here the same dynamics of the genoa cars come into play.

In fact, under a puff of 30 knots, these 1276 kg refer to the maximum working load applied to the purchase, because if the wind gets any stronger and we do not let go of the mainsheet, the boat will be knocked down and the load will come off the main. But the really important thing to note is that nobody will want to trim the main with wind this strong. In fact the opposite is true, and either we will ease the sheet or come up to wind to ease the load on the sails.

Well, while on the one hand we assume we can haul 34 kg and thus with a 30:1 purchase a theoretical 1020 kg, or even 918 actual kg taking into account a 10% loss of efficiency in the purchase, on the other hand we know that the difference between the 1276 kg maximum working load, the peak value, and the 918 kg maximum working load developed by the purchase is in no way a nonsense or a mistake in calculation.

The value of 1276 kg serves to allow us correctly to dimension the purchase or whatever other system we choose for trimming the mainsail; it is the maximum working load the system will have to support. In other words, it is the peak load under which the system must continue to work, in the sense that it must be able to handle this peak load without appreciable deformation, even if the system is not able to develop a pull of 1276 kg with the 34 kg applied at the 'entrance' by the trimmer, as we explained earlier. The 918 kg developed by the system with the 34 kg input are called the 'pulling load' and the maximum working load of 1276 kg is called the 'holding load'.

But just how do we dimension a system for trimming the mainsheet? The first thing to do, as we have seen with the genoa, is to formulate a value for the maximum working load on the mainsheet. There is a version of the Marshall formula for the mainsail:

$$\textbf{Working load in kg} = \frac{E^2 \times P^2 \times V^2 \times 0.02104}{\sqrt{(E^2 + P^2)} \times (E - X)}$$

[5] This is a delicate subject and has long been debated. Recently I met the crew of a Star whose bowman, weighing in at around 100 kg, a professional sailor who trained every day in the gym, repeatedly hauled, with two hands, about 80 kg. The 34 kg mentioned refer to the pull exerted with one hand by a reasonably fit non-professional sailor.

Where E is the length of the foot of the sail, P the length of the luff and X the distance between the end of the boom and the point where the mainsheet is attached[6] (all expressed in metres); and V the apparent wind speed in knots.

In fact, this version of the Marshall formula always leads us in practice considerably to overestimate the load on the mainsheet. A formula that gives more realistic values is:

Maximum working load in kg $= SA \times V^2 \times 0.02104 \times 1.5$

where SA is the area of the mainsail in square metres and V the speed of the apparent wind in knots.[7] So for a main of 45 m^2, at 30 knots we will have a maximum working load of 1276 kg. Given that this is the peak load that our system will have to bear up to, without suffering damage, we can go one of two ways. The first is to divide this load by the 34 kg that represent the maximum pull the trimmer can exert with one hand, and thus arrive at a value of 37.5, which we round down to 37. This is the number of parts that the purchase will need to do its job. We can obtain a purchase of 37, or rather 36, by using a cascade system consisting of two 6:1 purchases, but here the main defect of such a system quickly comes to light: the huge amount of line we would need to haul on board when, on a run, we need to bring the boom amidships because we want to gybe or sail close-hauled. On a 40 footer this would mean hauling in about 30 m of line!

It is easy to calculate how much line you need to haul in with a purchase to bring the boom in from its position with the mainsail completely eased (on a full run or a broad reach if you have a swept-back rig and the boom cannot go beyond that position): multiply the distance between the fully eased boom and the centre of your mainsheet system and multiply it by the number of parts in the purchase (if you have a cascade system only take into account the parts of the main purchase, since that is the one you will use to haul in the boom quickly).

Thing are not much better in the other direction: try easing 30 m of mainsheet in a hurry when you have to give way suddenly to another boat: you will soon realise how hard it is to keep almost 30 m in a shipshape state in the cockpit! In fact, it is no accident that modern 40 footers have almost all opted for a system with a 2:1 purchase that uses winches.

Or, if you really do want a block and tackle system, bear in mind that those 1276 kg are only the maximum load the system must be able to bear, and set a

[6] The higher the value of X the higher the load, for the same wind and sail area. Mainsails with the sheet fixed half-way down the boom develop very high loads.

[7] This simplified formula holds good as long as the value of 'X' does not exceed 20% of boom length; otherwise you need to use the earlier version.

new working load for the purchase, 20% lower than the earlier one of 1020 kg. Now, if we divide 1020 kg by 34 kg, we get a value of 30, which is the number of parts in a system we could construct with a 6 : 1 fine tuning purchase and a fast purchase of 5 : 1. At least in this way the amount of sheet to haul in is down to 25 m, with the boom right out.

I would like to underline the substantial and very important difference between the maximum supportable working load, the load a system has to be able to bear without undergoing permanent deformation, and the maximum operational working load under which the system not only does not suffer damage but can also do the tensioning job it was designed for.

If we want to avoid the problems that are created by using a tensioning system consisting solely of purchases, which is very light but has the defects described above, for boats with mainsails of around 40–45 m^2 we have to devise a system that includes one or more winches associated with a purchase. On boats with smaller mainsails it is fine to use purchases alone.

In the above case the purchase will allow us to reduce the initial effort but also, and above all, permit us to double up the mainsheet circuit so that we can use it both from the windward and the leeward sides (see Fig. 4.6).

The 1276 kg maximum working load on the mainsheet should, as a general rule, be divided by the number of blocks or parts[8] in the purchase we decide to use. If there are two blocks, as in the illustration, then each block must bear half of the expected load plus – note well! – the load increase factor relative to the angle at which the line is working. The block on the car of the main track will have a working load of 1276 kg, while each of the two blocks on the boom will bear half of this load, 638 kg, but this must be multiplied by 1.41, the load increase factor for a block that turns a line through 90°. So that means a working load of 899 kg for the blocks in the circuit that turn the line through a right-angle.

Now look at Fig. 4.7 that shows a 3 : 1 mainsheet purchase but with the sheet fixed to an eye on the boom and a block that turns it through 90° (though the inefficient position of the second block effectively reduces it to a 2.5 : 1 purchase). This system, though it is designed for blocks of the size found on dinghies like the Lightning, is a good illustration of the problem mentioned above. If the load created by the main is 50 kg, the block on the track must have a working load of 50 kg, while the block under the boom that turns the line through 90° must have a working load of $25 \times 1.41 = 35$ kg and the shackle on the eye only 25 kg, since the line is not turned.

[8] Do not be deceived by the illustration of the mainsheet system: it might seem that there are four parts coming from the boom, the mobile element, but in fact there are only two – those at the end of the boom – because the other two, those close to the gooseneck, are simply lead blocks on a part of the boom we can consider fixed, since the arc of movement at that point is very small. So it is a 2 : 1 circuit.

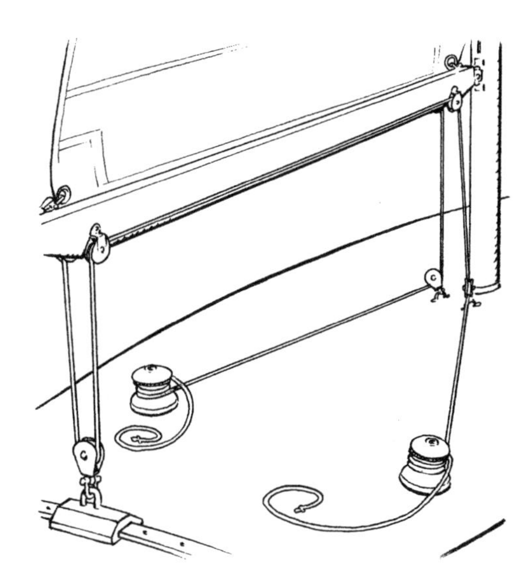

4.6
Mainsheet system with 2:1 purchase and a winch each side.

Another example. In the photograph of Fig. 4.8 you can see the same kind of system applied to the mainsail of a 46 footer, but here the eye and the block are attached to the same point, to the lashing done with the strop. And note that the strop must bear the sum of the two loads (the wear on the strop seems to support this theory!).

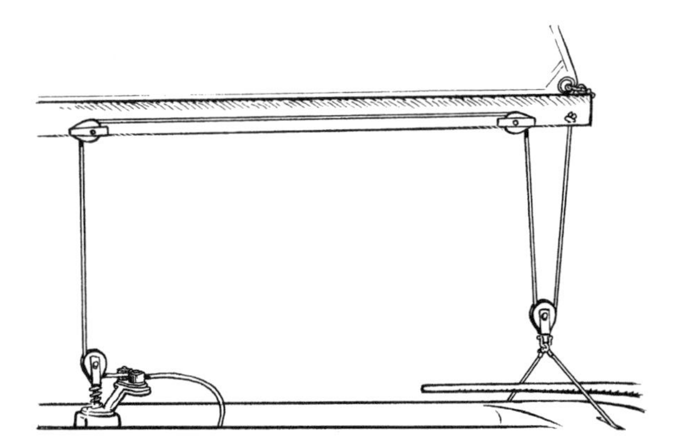

4.7
Mainsheet system for Lightnings with a rather inefficient 2:1 purchase.

On the car we will be careful to use a block that can stand upright to avoid it banging on the deck in light wind, so we will have to use a spring or special supports in tough rubber between the block and the car. Or we can choose a car that comes ready with a block that has these characteristics.

4.8
2 : 1 mainsheet system for a 46 foot cruiser with block and pad-eye attached to a strop around the boom.

4.9
Mainsheet system for a VOR 60 footer with a 2 : 1 purchase (and winch) with pad-eye and block at the end of the boom fixed to the same eye.

On the boom we will use single blocks with swivel shackles so that the blocks can rotate freely and assume the best orientation required by the sheet for different positions of the boom without suffering damaging twisting. The shackle on the block on the boom anyway implies the existence of an eye or other attachment on the boom, but it is preferable, for reasons both of lightness and simplicity and for efficiency in general, to use blocks to which lashings in spectra can be attached. These quickly align themselves with the direction of the load. On the deck we could use a single stand-up block, one whose shell is always held upright by a support, but these have the disadvantage of standing well proud from the deck, and this can be bothersome for the crew that will often tend to trip over it. It is better to avoid this problem by using a flip-flop block where the line exits low down, like the Frederiksen model in Fig. 4.10.

4.10
Frederiksen block for a 2:1 mainsheet fixed with a Spectra lashing for a VOR 60 footer; below, a Harken single stand-up block with toggled attachment and, right, a Frederiksen flip-flop deck block with low line exit.

I want the concept of the load distribution according to the number of parts in the purchase assumed by the formula above to be clear. So I would ask you to make a comparison with the two illustrations overleaf. This purchase has just two blocks: one on the car and one on the boom. The one on the boom bears 100% of the load produced by the mainsail. But the purchase system we looked at was also 4 : 1, but had the load produced by the mainsail divided equally between the two blocks on the boom, so would need smaller blocks than the other system.

But let us go back a moment and dimension the 2 : 1 circuit we looked at, and choose the winches. We had a working load of 1276 kg which, divided by 2 since each of the two winches will support half the load, gives us 638 kg, which in turn has to be divided by 15 kg, the power exerted on the 10 inch winch handle. This equals 42, so the winch must have a power ratio of 42 : 1. Obviously, the accuracy of this calculation depends on the assumed value of 15 kg. If this is higher a smaller winch will do, if it drops we will need a bigger one.

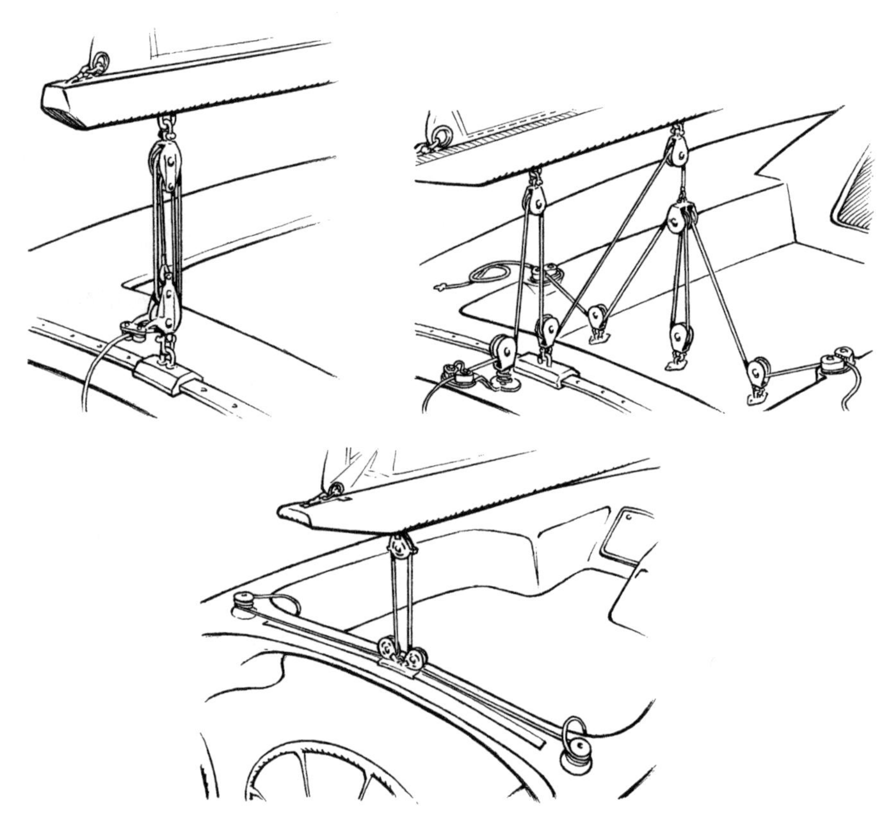

4.11

Left: a 4 : 1 mainsheet purchase; on the right the load of the mainsail clew is divided between the two blocks on the boom, the total purchase of the cascade is 16 : 1; below: mainsheet system with a 2 : 1 purchase, an athwartships layout with two winches.

To complete the picture, let us not forget that the 638 kg assumed as the load on one of the two parts of the purchase will be increased by about 2% for each turn of the sheet through 90°, even if the blocks that do this have ball bearings, and that the efficiency of the winch will be about 70%. So instead of exerting the pull of 630 kg given by the 15 kg input multiplied by 42, the power ratio of the winch, it will in practice exert a pull of 441 kg. So we will either have to increase the 15 kg manpower input or use a bigger winch!

4.12
2 : 1 purchase system on a 30 m maxi cruising yacht.

The variation of the 2 : 1 mainsheet circuit we described earlier and illustrated in Fig. 4.12 looks very similar, or even better, for it uses far fewer blocks and both the amount of friction generated and the weight and cost of the system must be lower. The two latter points are true, but while it is true in general terms that the friction is lower, if we talk specifically about the lateral friction caused by the sheet that has to run horizontally in the two blocks on the car we will be wrong. In fact this system, which was used on very many IOR boats in the 1980s, is very compact but has the defect of having a very high lateral load on the car, caused by the two horizontal parts of the sheet leading to the winches. To ease the car to leeward or haul it to windward you have to overcome the friction the sheet produces on the two blocks on the car.

Okay, but how much force do you need to exert to bring the car to windward? Bearing Fig. 4.13 in mind, if the load on the car is 1000 kg, the force needed to move the car to windward will be 25% of that load, or 250 kg.

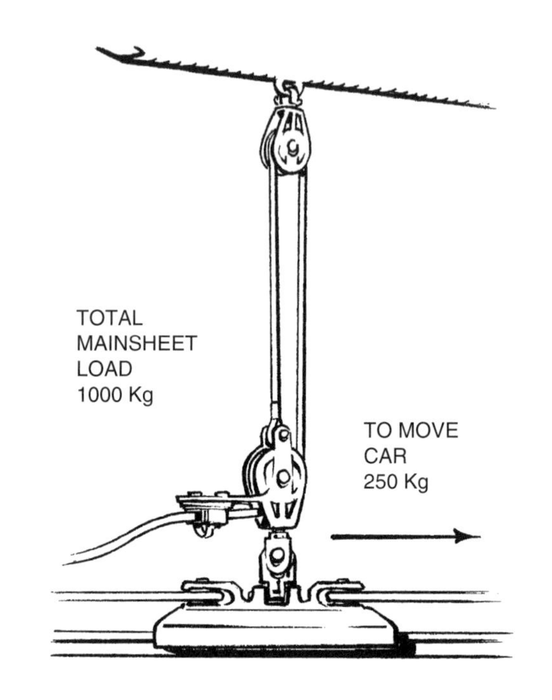

4.13
The force needed to bring the traveller car to windward.

Note that this is true if the circuit of the mainsheet is laid out in a fore-and-aft direction. But if we have a circuit with an athwartships layout, like the one we are considering, we will need a force of as much as 40% of the total load on the car to move it.

A practical example? Look at the photograph in Fig. 4.14: the plate that supports the sheaves for moving the car has been literally buckled by the load.

Another example: a fractionally rigged 90 foot maxi yacht had this kind of system for its 200 m^2 mainsail and with our 30 knots of apparent wind this main will put a load of 5670 kg on the car on the track. Forty per cent of this load is equal to 2268 kg. Originally the car was moved by a winch with a power ratio of 48 : 1 using a 2 : 1 purchase, so the winch bore a load of 1156 kg (2268 kg divided by 2 but increased by 2% because of the friction). This made moving the car a very tiring business. Changing to a 3 : 1 purchase, which put a load of only 771 kg on the winch, made things a lot better.

Another not indifferent problem with this system comes from the fact that both the main sheet and the car control line have to be led athwartships, so it is often hard for the mainsheet to reach the winch or the block that turns it if the winch is not perfectly in line with the end of the track without rubbing over the end stops used for controlling the car.

4.14

Above: The plate that holds the blocks has buckled. Below: a 2 : 1 mainsheet system on a 30 m cruising maxi with athwartships layout.

This is why – mistakenly – people decide to guide the mainsheet out of line with the track, and here the problems start: look at the photograph (Fig. 4.15) showing the position of the car when it reaches the end of the track. You can see clearly that despite the flat block with its extra-large sheave, the mainsheet tends to chafe against the upper cheek of the flat block. This is because as the car moves along the track the angle with which the sheet leaves the block on the car changes. So what is needed is a solution like the one in the illustration in Fig. 4.15. This avoids all the problems, since not only is the block in line with the track, but it is also self-aligning.

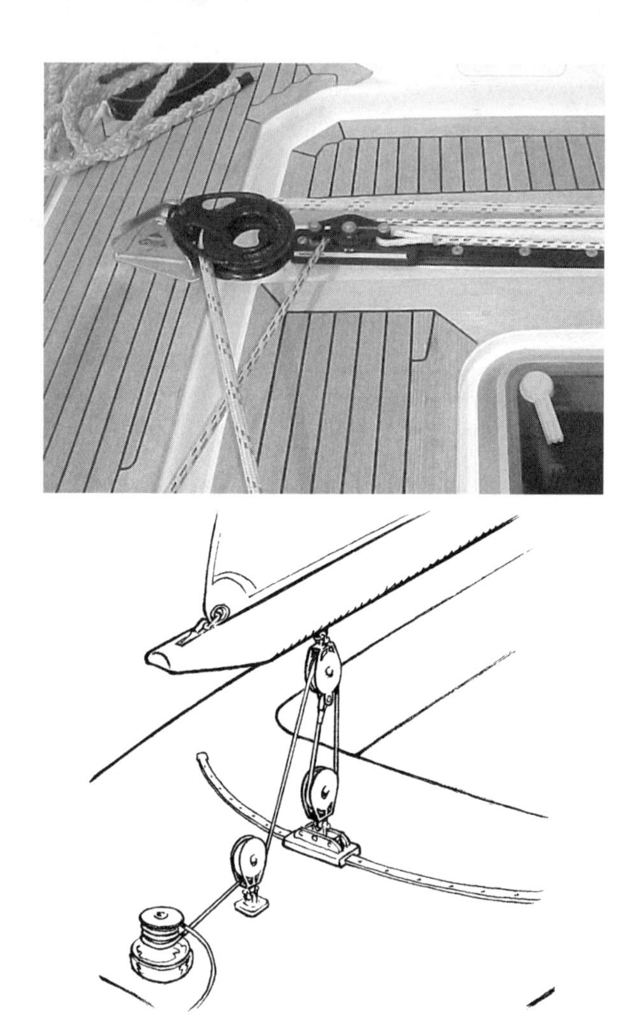

4.15

Flat self-aligning lead block for a 2 : 1 transverse mainsheet system on a 60 foot cruiser; below: a 3 : 1 fore-aft mainsheet system.

To sum up: avoid, as far as possible and if the deck layout permits it, a circuit for the mainsheet with an athwartships layout; fore-aft layouts are highly recommended for the reasons we now know. If you simply have to use an athwartships layout, try to make sure the block turning the mainsheet runs freely as we described, and that the system for the car control line is sufficiently powerful. In truth it has to be said that systems with a fore-aft layout mean that the point where the sheet is trimmed and made fast is rather a long way from the rest of the purchase, and in some deck layouts this can be a problem. A widely used system on the first wave of maxi yachts, such as *Kialoa, Helisara, Gitana,* etc., shown in Fig. 4.15, had a fore-aft type system but still with the handiness of having the point of trimming and making fast close to the rest of the purchase. But it is still worth remembering that with this system, where the total load of the mainsail is divided – on deck – between two points (the stand-up block and the car on the track), once the car is dropped to leeward the sheet, since it is rove through the stand-up block amidships, always tends to hamper the leeward movement of the car.

We can see this lateral resistance component very clearly in a kind of inverted 'W' mainsheet circuit used on many of Peterson's IOR boats. This system has been unjustly forgotten by today's boat designers, I believe because of the problem we are now talking about, but it does have two big advantages. The first is that, at deck level, the working load is spread over three separate points: the car on the track and the two blocks. This means a car with a much lower working load can be used (and that means a car that is less costly, less bulky and less expensive) compared with a system like the inverted 'T', where the car has to bear all the load on the deck. The second advantage is that, despite the athwartships layout, the car itself only bears a vertical load, and is thus much easier to haul to windward or drop to leeward. It is certainly a system that deserves to come back in fashion.

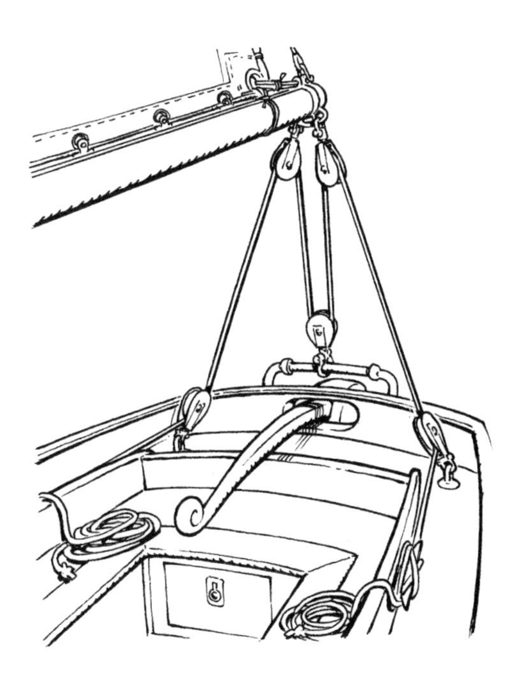

4.16
Mainsheet system with an inverted 'W' 4:1 purchase.

In the Marshall formula for the mainsail, you will have noted the presence of a factor called 'X', which is the distance between the end of the boom and the point where the mainsheet is attached. The greater this distance, the greater the working load, for the simple reason that the boom acts as a lever, and so the closer to the gooseneck the sheet is attached, the shorter the lever and the less advantage it gives us. So we have to work harder! Despite this, however, on cruisers, and especially on those designed with extensive charter use in mind, the track for the main is almost inevitably absent, or at least relegated to the roof of the cabin, so it is not in everyone's way in the cockpit. The illustration below gives a good idea of this solution that is a classic on the kind of boats we are talking about.

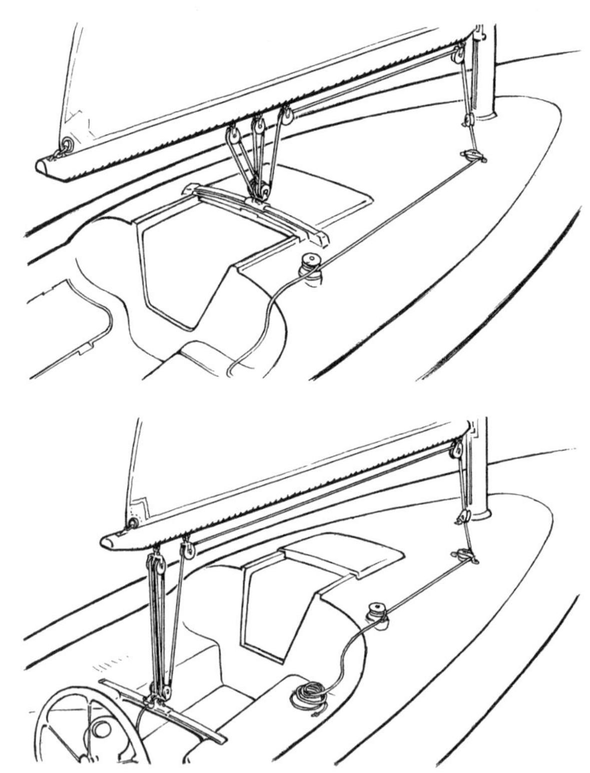

4.17

Mainsheet system with traveller on the cabin roof; below: a mainsheet system with the traveller in the after part of the cockpit, fast hauling with a purchase and fine tuning with a winch.

The illustration above shows how much of an encumbrance the track would be in the middle of the cockpit. Note, by the way, the unusual solution adopted of having fast trimming using the purchase on the track and fine tuning using a winch on the cabin roof, which can perfectly well be a winch used for a halyard.

Speaking of the mainsheet system with the track passing over the hatch on the cabin roof, you may have noticed that the track is 'suspended', that is, it is not fixed to the cabin roof along its entire length. This is possible either by using a special track riser made by a number of deck equipment companies, or by using special 'self-supporting' tracks. Obviously there is a limit to everything, so with all kinds of track you must be careful about how much of its length may be left not bolted to the deck.

For example, the 32 mm wide Harken track for cars with ball bearings, suitable for offshore boats from 36 feet up, can have a 'suspended' length that is not bolted to the deck of 1250 mm with a mainsail of 51 m^2.

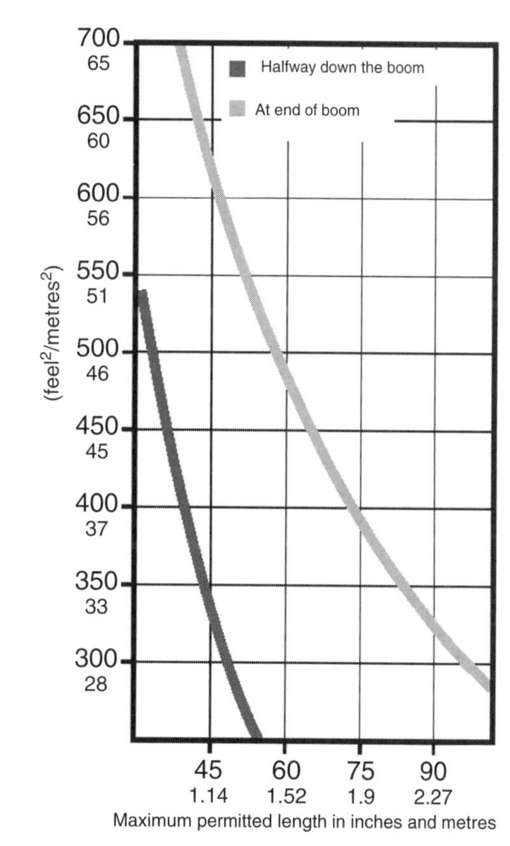

4.18

Graph for Harken raised track showing the maximum permitted unsupported length plotted against the sail area (note that the black curve refers to a mainsheet attached half way down the boom, which means a shorter unsupported length for a given sail area).

For a 33 m^2 main with its sheet attached half way down the boom the track may remain unsupported for a length of 1.14 m.

To complete the picture, you can have mainsheet tracks, like genoa tracks, that come already curved or can be curved during installation. Before ordering these, check with the supplier that the kind of track you intend to use can be curved as you require (give the length and chord depth) without being weakened or suffering anodisation damage.

One final point. We have spoken of mainsheet systems based on purchases and winches, when in fact back in the IOR days it was very common to see boats with a third element present in their mainsheet system: a hydraulic cylinder. In fact Tim Stearn, who was the father of masts with three sets of spreaders at the start of the 1970s, extended his influence with hydraulic systems consisting of a central manual pump with a series of valves and manometers that operated one hydraulic cylinder for every function required.

This mania for hydraulics got to the point where in some years you could alter the rake[9] of the mast while racing, because the forestay was connected to a hydraulic cylinder. In later years class rules forbade this, and at the same time the overall need for less weight on board considerably limited the use of hydraulics. However, you often saw a 2:1 purchase with an athwartships layout with a winch each side, and the block under the boom connected by a flexible wire rope to a hydraulic cylinder inside the boom for fine tuning. Once the mainsheet had been trimmed with the cylinder fully open, with its piston fully extended, you could fine tune by pumping the handle of the hydraulic cylinder. In this case the cylinder

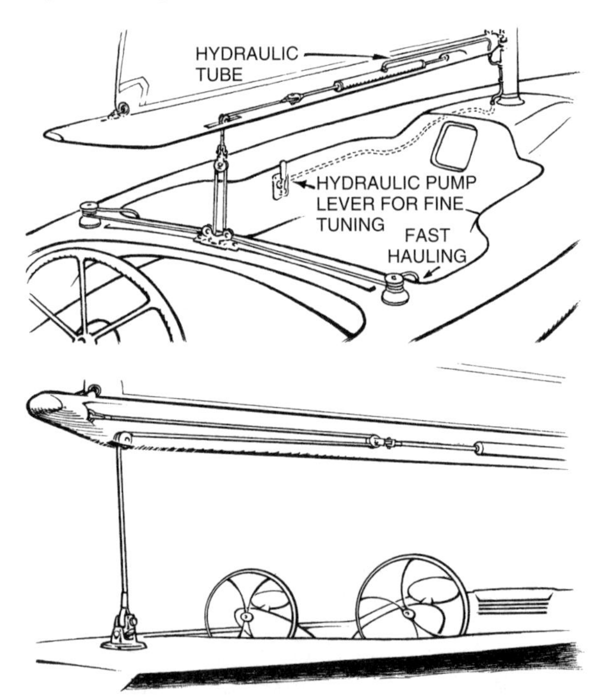

4.19
A 2:1 inverted 'T' purchase with a hydraulic cylinder for fine tuning; below: mainsheet with a hydraulically driven inverted 2:1 purchase.

[9] The rake of a mast is its inclination towards the stern. When sailing close-hauled it is a good idea to have the mast well raked for aerodynamic reasons and because moving the main aft obviously gives us weather helm, while on a run it is best to reduce the rake to almost nothing to have the spinnaker as high as possible. Rake should not be confused with pre-bend, which is created by wedges placed in the step of the mast.

acted like a reverse purchase, in the sense that instead of acting as purchases usually do on one part at a time, the cylinder acted, thanks to its load capacity, directly on a block that was able to tension all the parts at once. Obviously the cylinder needed to be dimensioned for the entire load created by the mainsail plus the extra load due to friction in the system.

The alert reader may ask the reason for the apparently needless complication of the block that runs inside the boom, since in any case the cylinder must bear the entire load. If the cylinder were connected directly to the mainsheet, it would need a stroke length equal to the maximum length it needed to haul in, which is obviously the distance between the point in which the boom is fully run out and when it is amidships. This means it would be impossible to use a system of this kind because a length greater than that of the boom itself would have to be recovered, and there would not be enough room for the cylinder. Of course the difference with respect to the system we described previously is that here the cylinder directly recovers as much as is necessary of the sheet itself, while in the other system the cylinder only recovers the final part of the sheet needed for fine tuning.

4.20
Caritech Magic-Trim hydraulic cylinder.

Here too the Wally yard, working with Caritech, a product of the Italian mast builder Cariboni, has made its mark on the rigging of yachts that use the easy-sailing philosophy with Magic-Trim. The truly brilliant idea was to use a particularly robust hydraulic cylinder[10] fitted, top and bottom, with twin sheave blocks through which is rove a line forming an (inverse) purchase with four parts. When this is tensioned as the cylinder expands, it recovers a full 4 m of sheet for every metre the cylinder is lengthened. That is not all. Since this inverse purchase is, so to speak, self-contained, all the load is internal to it and there is no stress on the structure of the yacht. Not bad, is it?

[10] Robust should be taken as meaning suitable for working not for pulling, as is normal with cylinders for marine use, but in compression, since when the cylinder expands to tension the line it is subjected to compression.

The use of the Magic-Trim permits incredible systems for recovering line which, even if they also have winches for safety reasons (always have back-up systems on board in case of breakages) and for speed of recovery, are so incredibly compact they can be hidden away inside booms or the sides of the cockpit. They can also be easily fitted with servomechanisms that do a lot to automate manoeuvres.

At the other extreme, on today's mega yachts, we find, paradoxically, an updated version of the hydraulic system: the hydraulically operated line storage winch. This is a hydraulic winch, generally sited below decks, with a very large and specially shaped drum, rather like a fishing reel, that allows it not just to tension the sheet but also to reel it in, easing it when needed. It is basically the fabric rope version of the old winch on which wire halyards were wound in. This kind of winch means you can have a very clean and uncluttered deck layout, since all you need, given the power of the winch, is a 2 : 1 purchase for the mainsheet or even a 1 : 1 system that goes directly to the winch. So it is easy to appreciate its validity for a cruiser where you do not want the deck cluttered with purchases and blocks. Its weak point is that the sheet must be carefully guided all along its route from above to below decks, otherwise its sleeve will be destroyed, and the line itself chafed to the point of laceration, and the turns on the drum below will foul.

We must point out one thing concerning the mainsheet systems present on modern mega yachts designed for 'easy sailing'. Here no mainsheet track is used, and instead there is a pad eye on deck (in the case of a direct 1 : 1 circuit) or a simple block in the cockpit (in the case of a 2 : 1); this same block may be at the apex of an inverted 'Y' or may be fitted to run along an adjustable loop of line, like the mainsheet of a *Laser*. All these systems need a boom that is specially designed and built to withstand the high load there will be on the vang, which will have to prevent the boom lifting when sailing downwind. So this kind of system needs to be studied and planned for before the boom is constructed. That is not all: you need to pay special attention to these 1 : 1 systems with no purchases in them. It is true that today's winches are increasingly problem free, but it is also true that our hands have not evolved with them. And finding yourself, at the mainsheet of a maxi yacht, holding a sheet that more often than not has a working load of more than 5000 kg is no laughing matter. If you take one turn too many or too soon off the winch, out goes the sheet and your hand goes with it!

Let us end this look at mainsheet systems with a circuit borrowed from none other than the mainsheet of the old *Vaurien*, which has the advantage of solving the problem present in all systems without a traveller. The problem these systems have is in bringing the boom amidships, since the sheet is attached amidships and there is nothing pulling on the boom from the windward side. This circuit does not make the traveller redundant but it does have a part that hauls from the windward side. Thus it can bring the boom amidships and the sheet also acts as a vang.

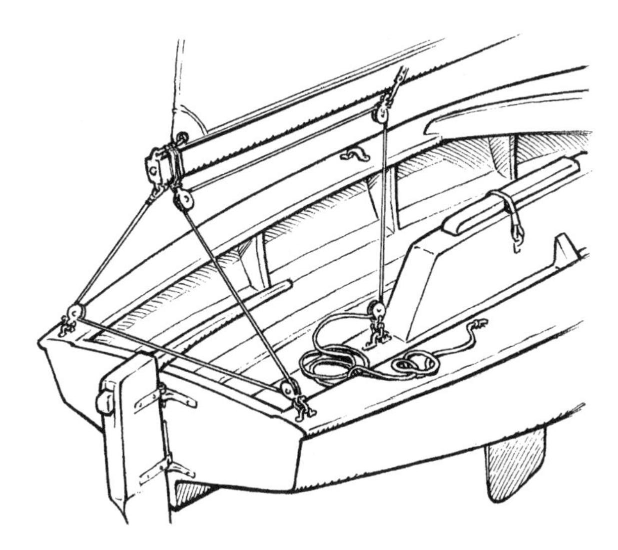

4.21
Vaurien type 2 : 1 mainsheet system without traveller.

A few words about a particular kind of mainsheet. In the second half of the 1980s – it was already the era of lines in Spectra – it used to make me smile to see old IOR glories from the beginning of the 1970s, two tonners such as *Iorana*, using mainsheets like the Intrepid: a 12 strand line in polyester with a 'hairy' finish that was very soft and well-balanced, in that it had half of its strands right-laid and the other half left-laid, which produced line that tended to have as few kinks as possible. It was made by an American firm, Columbia, which is no longer in existence.

In fact, it is true that the mainsheet too has followed the evolution in lines in recent years, but it is also true that it is a piece of rigging that does not leave much room for creativity. Let me explain. In the obvious case of the 2 : 1 system where there is no dead-end and either 'side' of the sheet can be the one that takes the maximum load, the only sensible thing you can do is to choose an excellent brand of line, decide what material it should be made of on the basis of how you use your boat, and determine what size you need. And that is about it. But on a mainsheet system with a dead-end, you can get up to some tricks, since you can determine precisely how much of the sheet will be under the maximum load (you should determine this with the boat sailing close hauled). Then you can have this part of the sheet in line that has the appropriate mechanical properties, then use a lighter line that will be suitable for reaching and broad reaching, then use a very light tail for running.

Finally, a brief look at how to calculate the length of the mainsheet. In a direct 1 : 1 system you will need a length equal to the distance between the boom fully run out and the point where the sheet is recovered in the cockpit, plus the distance between this point and the recovery mechanism, be this a winch, a hydraulic cylinder or a line storage winch. If on the other hand the sheet uses some kind of purchase, you will have to multiply the distance from the fully run-out boom to the centre of the cockpit by the number of parts in the purchase.

The outhaul, which allows us to adjust the foot of the mainsail, gives us an important control over the sail. Tensioning the outhaul flattens the lower part of the sail considerably and improves close hauled performance. A more bellied foot will help in light airs and when sailing off the wind.

A system for the outhaul must – as usual – satisfy two conflicting requirements. It needs to be powerful enough to allow us to tension the foot of the main in fresh wind, but will need to be released freely if the wind drops. It is clear that a purchase with a large number of parts will be very powerful (useful in strong winds) but it will also mean a lot of friction in light airs. The best compromise must have a sufficient number of parts to the purchase but keep friction as low as possible.

In the case of mainsails that have the foot that slides into a groove in the boom, it is very important that the boltrope and the slide that is usually found before the clew are lubricated with a suitable spray such as McLube, because they cause a large amount of friction. Modern mainsails, where the foot remains free, do not have this problem. Generally there is a tendency to underestimate the load on the foot of the sail, but in fact it is very high: bear in mind that if, as often happens, a car with ball bearings is used to allow the foot to run freely, this car must be able to bear a working load equal to that on the car of the mainsheet track. Another demonstration of this is that the strop[11] in Spectra that is passed round the boom in the case of mains with a free foot to serve as an attachment for the sheet is always quite large, comparable in diameter to the main halyard.

The table below gives an idea of the power ratio needed for the purchase on the outhaul.[12]

Table 4.2

sail area	purchase ratio
9 m^2	3 : 1
14 m^2	4 : 1
19 m^2	6 : 1
28 m^2	8 : 1
37 m^2	12 : 1

[11] This is a band of fabric, sometimes even tubular in section, generally in polyester but also in Spectra, which has traditionally replaced on modern boats the use of reef points fixed to the mainsail; instead lengths of fabric are preferred for lashing the reefed part of the sail. From this point the use of these fabric bands spread just about everywhere; they are used to fix the spinnaker pole to the stanchions, to lash a tender on deck and are even worn as a belt over foul weather gear.

[12] From 'A purchase for every mainsail foot' by Marty Rieck, in *Technical Sailing*, pages 6–9, n.1 January–April 1995.

Spinnaker sheets and afterguys

The spinnaker sheets must be one or two sizes smaller than the sheets of the number three genoa. In the mid-1980s, on top One Tonners it was usual to have for the sheet a 6 mm Kevlar line with a 10 mm tail spliced on to it, because this was the minimum diameter needed for handling the line. This tail in thicker line is pretty rare these days because you can get the same effect by removing the sleeve from a line in Spectra along the working part and leaving the sleeve on the 'tail' part. But at the time it was a true revolution, brought about by the English rigger Peter Morton.[1] In general it allows you to avoid carrying on board two types of spinnaker sheets, one for light breezes and the other for stronger winds, which was common practice on boats of 20 years ago. You can calculate the length needed for the spinnaker sheet by multiplying the length of the boat by 1.8.

The spinnaker afterguys have lived through the entire passage from three-strand tope to double braided line spliced to wire rope, right through to the exotic fibres. The afterguy on a beam reach is, together with the number three genoa sheet, one of the lines on board that is subjected to the highest load, together with the runner tails (on a fractional rig) and last but not least the code 0 tack or halyard: an authentic winch killer!

There is no Marshall-type formula for calculating the working load on the spinnaker clew. For the afterguy, it is excellent practice to use line equal to or one size larger than that of the number three genoa sheet. In the days of afterguys in line spliced to wire rope, it was hard work for a rigger to prepare a good afterguy because the final part in wire, which was rove through the end of the spinnaker pole, would fray and develop broken strands pretty quickly. So an expedient was used to reduce this. It consisted of doubling up the final half metre of the wire

[1] Peter Morton was the undisputed leader of the new approach to rigging in the 1980s. As head of the English Riggarna he brought an extremely innovative impulse to the sector. Those who came after him – in the vast majority of cases – were simply people who borrowed the ideas of others. This is all the more serious because the arrogance of certain people, especially on their home territory, has prevented them acknowledging their debt to Peter. Shameful.

rope, just before the shackle, with the same wire used for fitting the shackle: in practice, one ended up with an eye splice with a long eye. This meant having twice the thickness of wire, which lasted almost twice as long.

The afterguy was made with the wire rope spliced first to the shackle with a *nico-press* swage and then spliced by hand – wire to wire – and well tapered. The section was then hand whipped with *monel*[2] seizing wire to avoid this big eye snagging in the end of the pole.

At Spencer Rigging, I saw afterguys which, instead of the wire to wire splice, used special talurits that were tapered so as not to catch in the end of the pole when the afterguy was brought aft.[3] A Baltic 64 footer had some interesting custom afterguys made for an Atlantic crossing that were light to handle on deck. The all-fibre guys, in polyester, already present on board were very heavy and it was a job to use them in less than favourable conditions. We used (this was in 1989) as material Samson's Spectron 12, a single braid line with 12 strands of Spectra without an external sleeve. The Spectra was chemically waxed to keep it compact and stop it catching in pins, rings and the other classical points on board where lines chafe. The only modification we made was to fit a polyester sleeve starting from the point where the afterguy went on to the winch drum with the boat on a beam reach. On a run, thanks to the decidedly lower load, the naked Spectra, though it was slippery, could be kept on the winch with no problems.

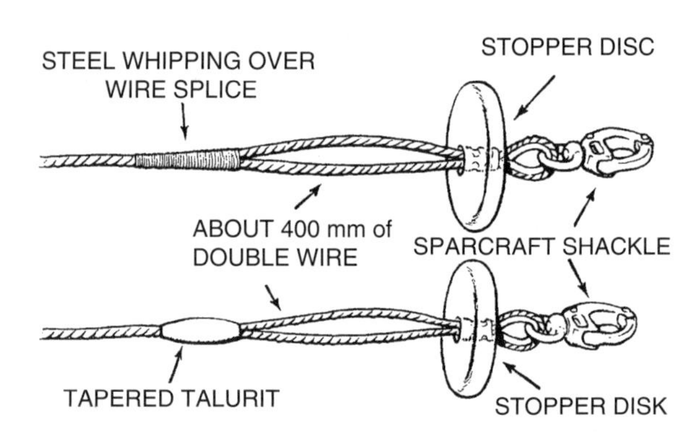

5.1

Spinnaker afterguys spliced with special tapered talurits.

At the end of the afterguys you must be careful to use a stopper so that the shackle does not jam in the end of the pole when the guy is trimmed: on the guys of smaller boats, of less than 30 feet, the classic PVC ball of the kind used for halyards

[2] *Monel* is a stainless steel that in the form of a thread is very useful because of its ductility for finishings or for use in place of pins. It is produced both by the English company Ormiston and the American Loos.

[3] Every time the boat is brought further off the wind, the pole must be brought further aft of the forestay by trimming the afterguy.

is sufficient, but as the boat in question gets bigger you must first use special disks, then those pear-shaped stoppers the Americans call *donuts*, shown in Fig. 5.2 (top). This illustration shows a Sparcraft shackle that was the first to allow the chute to be released under load, since it had a special mechanism for opening under load that was activated by a pin.[4] On all-fibre afterguys in exotic material, it was customary to cover the final half metre that was rove through the end of the pole in leather, but more recently in a brown tape in Teflon made by 3M that is very expensive but exceptional for preventing early wear on the sleeve. You need to be careful about choosing exotic fibres indiscriminately for all pieces of rigging. A classic case is that of certain single-handed sailors who avoid the use of exotic fibres precisely in the afterguys, because these fibres are so rigid and little subject to stretch that even the

5.2
Spinnaker afterguys in Samson Spectron 12. Below: an example of the use of steel stoppers on the afterguy.

[4] The Sparcraft shackle was made in 17-4-ph in the 6 and 10 models, respectively with working loads of 6000 and 10000 lbs, then came other models with larger and smaller working loads: they are easy to open with the finger of one hand so long as they are not under load. Today similar shackles are made by Gibb – also in titanium – and by Tylaska and Wichard.

slightest wind shift makes the autopilot work all the time. In this case it is better to use polyester for the afterguys, since this material accommodates small wind shifts by stretching and does not overload the autopilot.

5.3

Gibb Super-snap shackles that can be opened under load; below: pear-shaped 'donut' disc on the afterguy of a VOR 60 footer.

Usually the afterguy works directly 1 : 1, though some of the latest boats to sail around the world have used a 2 : 1 purchase because of the enormous forces a 1 : 1 afterguy exerted on the primary winches. This brought old methods back into fashion. As far as the overall length of the afterguy is concerned, it is a good idea to make this equal to the length of the boat in metres or in feet multiplied by 1.7. If you want a particularly lightweight and free running guy, though you will not be able to invert it, you can use a line suited to the working load of a length equal to the distance between the end of the pole – with the boat on a beam reach – and the primary winch, then splice on a tail in a typical lesser line. Watch out for the block of the afterguy: remember the increase in load due to the angle of deflection and do your sums carefully!

Halyards and reef lines

False Cleat for Mixed Wire Rope/Fibre Halyards

Storage Winches for Halyards Entirely in Wire Rope

Mixed Halyards in Wire Rope and Fibre

Halyard Tensioning Systems

Types of Halyard Lead Blocks

How to Lead the Halyards to Winches on the Cabin Roof

Twilight, a 44 footer designed by Scott Kaufman, had a curious gadget about half a metre from the main halyard winch: it was a kind of half cleat in metal, bolted to the deck. The purpose of this half cleat was to accommodate the excess of wire rope once the mainsail was fully hoisted.

This excess length was coiled[1] between the main halyard winch and the cleat in question (once this had been done, the tail of the fibre part of the halyard was made fast to a cleat). The 'excess' wire rope was needed when reefs were taken, so that the entire length, from the peak of the reefed main to the winch, was in wire and stretch was thus reduced to a minimum. While this was still relatively practical with the halyard of a 44 footer, when the same system was used on a maxi yacht like *Condor*, things got a bit trickier.

Uncoiling something like 30 m of 10 mm wire rope and handling it on a winch cannot have been easy . . . it would have been easier to charm snakes! But

[1] To 'coil' means to gather up a line in loops which may pass around an object. It would be a very safe bet that a modern sailing novice would be unlikely to know how to coil a spinnaker afterguy in wire and fibre: you have to start from the centre where the splice is and work towards the shackle, taking one turn clockwise and one anti-clockwise!

was there a valid alternative to a halyard made by splicing a supple wire rope to a fibre line, seeing that this was the solution adopted by racing yachts of the period? The answer is, absolutely not. The only alternative to the mixed halyard in wire and fibre line was one entirely in wire rope, all along its length: this wire rope was wound onto a special winch called a storage winch.

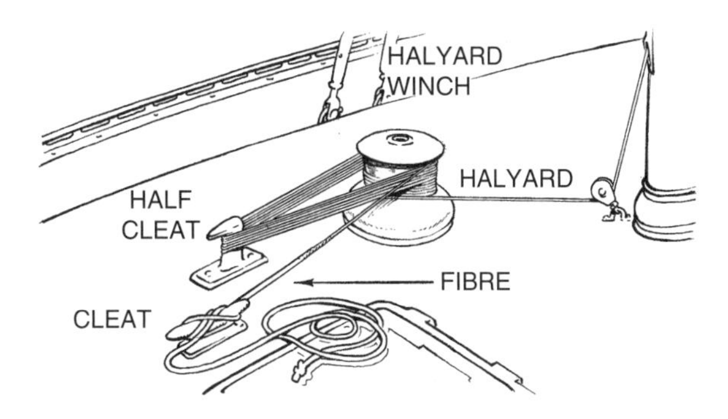

6.1

This kind of solution was mostly adopted by cruising yachts, since it offered the clear advantage of not leaving lengths of halyard around after the sail was hoisted: all the wire was wound around the drum of the storage winch. One the hoist was completed, you had to be careful to tighten a knob on the base of the winch, because otherwise, if the winch handle was left free to turn and there was no brake acting on the drum, the winch would start rotating in the opposite direction, spilling out the halyard and – something extremely dangerous – make the handle spin at high speed.

Another defect that was typical of this solution was the ease with which the wire rope took awkward turns around the drum as it rotated, adding new turns over those already on the drum. You had to be especially careful to avoid this otherwise you risked not being able to unwind the halyard. On the other hand, we must not forget that a halyard made entirely in wire rope that ends up on a storage winch mounted on the mast is an economical solution and a very tidy one from the point of view of deck organisation.

For boats that set sail not just to take a turn around the bay but for longer voyages, the disadvantages may not have faded into the background but were at least mitigated by the positive aspects described. Why was the same system not used on racing yachts? Because it was so slow! Even with the halyard winch placed on the deck and with a man at the mast to haul the halyard rapidly by hand (and this meant having the halyard leaving the mast about 3 m above the deck), the man at the winch who had to recover the halyard just by using the winch handle was inevitably unable to keep up with the man at the mast! The only way it could work was if the crewman at the winch hauled in the halyard by hand with just a couple of turns around the winch, and this was what led to the introduction of the mixed halyard in wire rope and fibre line.

And this illustrates the advantages of the two systems. Today, mixed halyards have almost disappeared from rigs and have been replaced by halyards entirely in fibre. It is easy to see why: stainless steel rope did have the advantage of having a very high breaking load for its small diameter, but had the big defect of needing a very large diameter sheave in the blocks it was rove through, otherwise it was quick to present broken strands that poked out from the wire and were a danger both to the crew that had to handle the line and to the sails, which risked being torn. A way of reducing to a minimum this tendency of the strands to break was to use galvanised steel rope. This is more resistant than its stainless steel equivalent because it is more supple and the strands have less tendency to break, but it does demand a lot of maintenance to stop it rusting. Angelo Vianello, the legendary mariner on board Raul Gardini's many yachts named *Moro di Venezia*, was a master of this art. He treated galvanised steel rope with a special mixture of linseed oil and petrol to delay the onset of rust for as long as possible. His halyards lasted longer than anybody's . . .

But today, of course, there is no time for anything, and certainly none for maintenance (which, it must be said, is the symbol of civilisation!) Where there is no maintenance, there is no civilisation. And this is true regardless of the greater or lesser technological content of the element in question: even the hut of a bushman requires maintenance, otherwise it goes to ruin. A steel rope, be it stainless or galvanised, has another big defect, which has become more significant over the years as more and more owners have wanted their masts painted and no longer anodised: wire rope plays havoc with their beautiful (and very costly) paintwork. And that is not all. When you slept on board, if you did not keep the halyards clear of the mast, the wire would slap against it and, in high winds, make a devil of a noise, while fibre halyards are much quieter. In any case, the price of exotic fibre line has dropped as it is becoming more popular.

Riggers themselves have gone happily along with the trend to all-fibre halyards, not just for the obvious reasons of extra turnover, but above all because it is much easier to prepare a halyard all in fibre line than one in fibre and wire rope. Then again, you can always reverse[2] a fibre halyard with no problems if one end of it is damaged, while on a mixed halyard you are forced to replace the damaged part, either the wire or the fibre line. To sum up, today a halyard in wire rope and fibre line is the most economical solution in terms of price-quality

[2] 'Reversing' means turning a halyard, or other piece of rigging, the other way around so that the part that previously bore no load now takes the load and the other part 'rests'. In other words, to reverse a halyard you cut the shackle free from its eye splice, or undo the knot to free it, then knot or splice it to the other end of the halyard.

performance, but it does have the disadvantages we described above. But if your wallet allows it, you can get an improvement in performance by using an all-fibre halyard.

The attentive reader will have noticed that the subject of this chapter – halyards – has been tackled by speaking mainly of the greater or lesser practicality of different kinds of halyards. But in fact we have deliberately glossed over an aspect that is fundamental for all pieces of rigging: the dimensioning. How 'big' does this halyard on my boat need to be? What breaking strain should it have? We can answer this question in two ways. One is by resorting to experience and a direct comparison with boats similar to our own in terms of length overall, displacement, rig and sail area, and thus try to take account of what has been done in the past.

The second, more theoretical method is to try to find out what will be the maximum working load on the halyard in question. There is no special formula for calculating this. The only formula we have is the Marshall formula in its genoa and mainsail versions, which gave us the working load on the clew. And we can find the working load on the head of the mainsail or genoa simply by reducing the result of the Marshall formula by 20%. Thus, if we have a working load of 1500 kg on the clew of the number three genoa, the working load on the head will be 1200 kg.

Circuits for the halyard? Nine times out of ten the halyard works directly, 1 : 1, because there is already a lot of line to deal with and if we add in a 2 : 1, or worse still a 3 : 1 purchase, this length will double or triple. But there are cases where a 2 : 1 or a 3 : 1 purchase is obligatory, that is when we use an asymmetrical code zero[3] spinnaker, because these are handled by being wrapped around a special kind of grooveless furler, similar to those used on FDs or Solings. Furling the sail in this way demands a lot of luff tension, which can be achieved only with a purchase of at least 2 : 1. If you opt for a 2 : 1, you will need a becket at the masthead and a block attached to the head of the sail, while with a 3 : 1 the becket will be on the block at the head of the sail.

It is funny to think that this seemingly modern innovation of halyards with purchases is only in fact a revisiting of an ancient usage common on classic boats where halyards more often than not used purchases.

Speaking of masthead contrivances for halyards, there is also the so-called 'lock',[4] which was very common on the International 12 m of the old America's

[3] A code zero is an asymmetrical sail – spinnakers are symmetrical because tack and clew are interchangeable – half way between a spinnaker and a genoa. The tack is fixed with a strop at the bow or on a bowsprit, or on the spinnaker pole held very low on the deck; they are not hanked on and reach almost to the stern.

[4] Actually the *lock* has been around for ages: old wooden masts of Finns, Yole OKs and Stars had a swallowtail-shaped piece of metal at the masthead in which the talurits of the halyards were blocked.

Cup, which allowed the halyards to be made fast at the masthead once the sails were hoisted. This system has the enormous advantage of removing from the mast the compression produced by halyards under load, and as a result making it possible to use very light halyard tails since these will only be needed to hoist the sail as far as the masthead, because at that point the lock took over the load. Someone might be wondering how the halyard was tensioned. Easy: with a downhaul attached to the Cunningham hole that tensioned the luff.

As far as the kind of line or wire to use for halyards is concerned, it is clear that although the working load is 20% less than that on the clew, the greater length of line under load compared with the sheet will lead us to choose the same diameter of line as that of the number three genoa sheet, so that the stretch will be the same in both cases. One we have dimensioned the number three genoa halyard, we will use the same line for all the other headsail halyards, while the mainsail halyard will be smaller if the rig is masthead or larger if it is fractional.

It is very important to have the halyards of as small a diameter as possible, so that the sheaves and their casings can be smaller thus avoiding large holes in the walls of the mast.

Concerning deck layout, and the choice of having halyards terminating at the foot of the mast, this will depend on the stability of the boat. On Sparkman & Stephens's *Dorade* it was no problem to have the winches at mast level, because leaving the cockpit to attend to the halyards on such a docile boat was a real pleasure. But on a modern boat, that is nervous and sudden in her movements, it is almost obligatory, as well as prudent, to have the halyards led aft into the cockpit.

Once again, pay attention, if you decide to lead the halyards aft from the foot of the mast, to the angle of deflection of 90°; this will increase the load on the halyard lead block by a factor of 1.41. Usually the blocks used at the foot of the mast are normal single stand-up blocks that have the disadvantage of making the line come out too high to be led easily to a winch or footblock. It would be much better to use low-lead fixed blocks for the halyards, though these need to be very carefully positioned so as to avoid incorrect alignment.

Once the halyards have been led away from the foot of the mast they can go directly to the winches or be turned by multi-sheave footblocks called organisers, if the hatch gets in the way of the former solution. In most cases it will not be possible to run the halyards directly to the winch because the hatch or something else usually prevents it, so organisers will have to be used. There are various systems used for tensioning the halyards, including 'magic box' type purchases or simple small purchases for dinghies or, on larger boats, winches, rigging screws, hydraulic cylinders or inverse purchases, the latter three mounted directly on the mast.

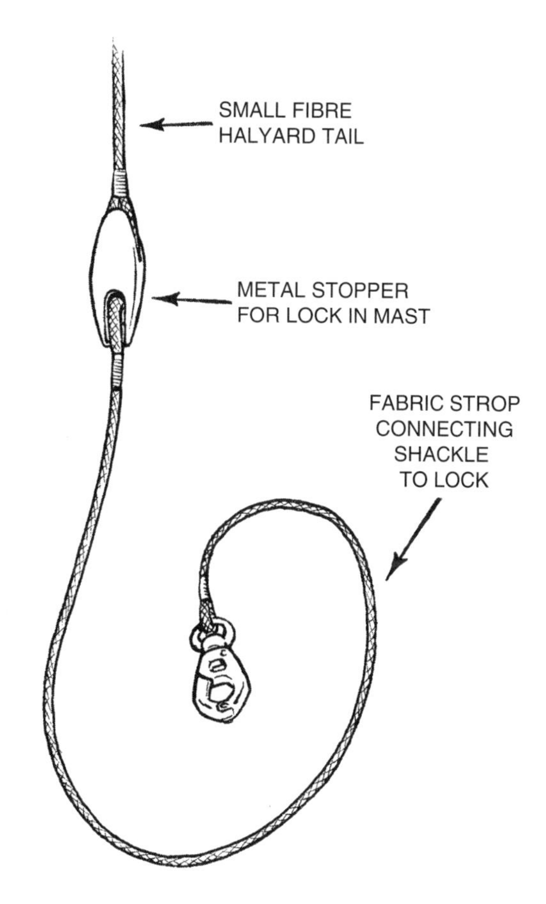

6.2
Lock system applied to the all-fibre halyard of a 60 foot cruiser/racer.

The tensioning system designed for Antal is very useful on big cruising boats where both main and foresails are roller reefed and thus not reefed in the vertical sense but horizontally, and where it is best to avoid having kilometres of line that can fall off the winches once the sails are hoisted. This system consists of a 'T' track fitted to the mast right under the exit point of the halyard and a car that runs along this track. Once the sail is hoisted, using a detachable tail, the end of the halyard is made fast to a cleat on this car. The car is then hauled down using a line leading from the cleat to a winch. Once the correct tension is achieved, the car is blocked in position by releasing the fixing pin, and the strop and tail are removed.

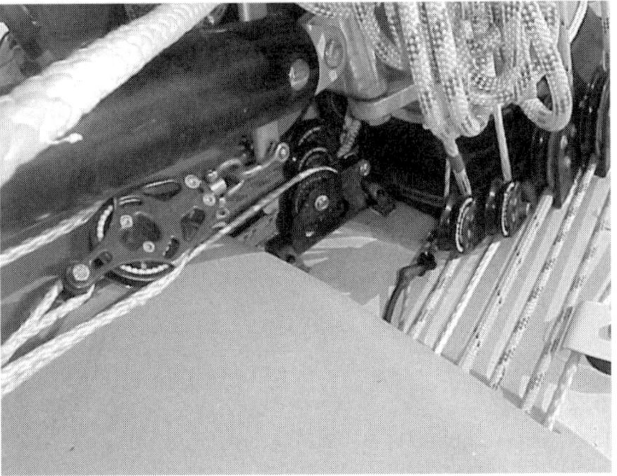

6.3
A 'low-lead' halyard block; below: mast step with 'low-lead' blocks that lead the halyards directly to the winch stoppers.

Obviously all the halyard tensioning systems that use mechanisms mounted on the mast will not allow a crewman at the mast to tension the halyard by hauling it away from the mast like a bowstring, as happens when the halyard is led from the foot of the mast to a tensioning system sited some distance away. This is the price to pay for having a very compact and neat system.

To establish the length of a halyard, though it is not accurate to the nearest centimetre, there is a simple rule: the length of the halyard in metres should be equal to the length overall of the boat in feet. For example, how long should the halyard of a 36 footer be? 36 m! Try it and see.

A reef line has exactly the same load as the leech of the mainsail, so we can use the Marshall formula for calculating the load on the clew of the main. But we

6.4
Stand-up block for reef lines.

must bear in mind that the reef line works as a bight, so this load should be halved. Let us look at an example: with a load of 2000 kg on the clew of the main, the load on each part of the reef line is 1000 kg, but be careful with the block the reef line is rove through at the foot of the mast, because this will normally be subjected to a load of $1.41 \times 1000 = 1410$ kg. Make sure you always use perfectly swivelling stand-up blocks (not fixed blocks), because the boom as it moves describes the arc of a circle close to the gooseneck where the reef lines exit.

Standing rig

Let us start with a curiosity: 'In the eighteenth century and up to the end of the nineteenth, there were several rules establishing the size and number of the shrouds that were dictated by experience more than anything. For example, the circumference of the largest shroud should correspond to $\frac{1}{52}$nd of the ship's beam.[1]

Today, however, we use sophisticated calculations with finite element programmes for racing boats, while for cruising yachts, with displacements that are known and are comparable with those of boats in the past, we know of what dimension the shrouds should be. For example, you can be sure that a 40 footer will have, 9 times out of 10, a forestay 10 mm in diameter if it is in wire, while if it is in rod the diameter[2] will be 8.4 mm, also known as '–17'[3] ('dash seventeen').

Some authors go on at length on this subject, trying to instil in the reader a kind of calculation methodology for dimensioning the shrouds. But this is somewhat suspect when you consider that not even the most expert mast builders always get things right, as demonstrated by some sensational dismastings by yachts that had been able to afford top rig geometry designers. And then it is unlikely that the average owner will be called upon to dimension the shrouds of his boat. Even if the boat is delivered without her shrouds, in 99% of cases she will have chainplates already fitted on deck, and the holes drilled in these will clearly indicate the dimensions of the shrouds needed. If your boat is brand new, let the mast builder you have chosen do the sums. In this field, DIY is dangerous!

What it is important to check – and this is something anyone can do – is whether the shrouds you have on board have the right diameter for the chainplates. This practical rule derives from the ancient practice of bringing sailing ships close inshore and heeling them as far as possible with a line leading from the masthead to

[1] O. Curti, *Il libro completo dell'attrezzatura navale*, Mursia, Milan, p. 118, 1979.

[2] If with wire rope – for a boat of this size – a diameter of 10 is almost a certainty, with rod the diameter tends to vary from the 8.4 mm mentioned to 8 mm. One of Riggarna's many merits was to offer 'round' diameters such as 8 mm, while Navtec always insisted on offering few diameters that were often typically Anglo-Saxon measures such as 8.4 mm or 12.7 mm.

[3] The Anglo-Saxons, with their great practicality, describe the various sizes of rod not by the external diameter but by a number preceded by a dash, as in 'dash 17 rod', also written '–17'. The number gives an approximation of the rod's breaking load in thousands of pounds. Thus '–17' rod has a breaking load of 17 000 lbs.

the quayside or shore so as to clean their bottoms. This put maximum stress on the series of chainplates[4] on the ship's side; they had to be strong enough to allow the ship to be heeled into an almost horizontal position. In more recent times, this practice has been echoed in the empirical method used to determine the righting moment of a yacht. This is done by applying maximum heel to the craft by hauling her mast down towards the quayside with a rope leading from the masthead and connected to a tension gauge. The greater the force needed to heel the yacht, the higher her righting moment. And this means that, in a given wind, she will be able to carry more sail than a boat with a lower righting moment.

7.1
A practical test to determine the righting moment.

These days a comparative analysis of the righting moments of various boats is a valid help in determining what deck equipment to use. Basically, a stiffer boat, with a high righting moment, will produce higher loads than a more tender one

[4] In fact, well-dimensioned chainplates should be at least 30% stronger than the shrouds fixed to them, for the simple reason that shrouds will be replaced occasionally while chainplates have to last the life of the boat.

that has a lower righting moment. It is for this reason that ULDBs, with their very light displacement and narrow beam, have lower loads on their equipment than boats of the same length overall, but with greater displacement and normal beams. The ULDBs have a righting moment much lower than that of the latter craft. Take too the 60 foot VORs used in the round the world race. With their water ballast tanks, they produce loads worthy of an 80 foot maxi yacht, for their righting moment is almost as high!

While we are on the subject, let us dismantle an all too facile criticism of the Marshall formula we saw earlier: that its weak point is that it does not take account of the righting moment. In fact the formula does, in its own way, take it into account: defining the maximum apparent wind in which we can use a given sail lets in through the window what we thought had been left outside the front door! In fact, the main of a ULDB will have to be reefed long before that of an IMS or IOR yacht of the same length in the same wind, otherwise the ULDB would be excessively heeled. Thus the formula, using the reduced area of the reefed main, will indicate a smaller load. So it does actually take account of the righting moment.

Another example is the *Hobie Cat* catamaran. This has a mainsail with an area similar to that of a single hulled dinghy, but uses for the mainsheet a purchase with almost double the number of parts a dinghy has. Why? Because of the righting moment! The *Hobie Cat*, being a catamaran, has remarkable lateral stability and a high righting moment, and can carry her mainsail in a much higher apparent wind than can a single-hull dinghy. And so the formula, which contains the square of the apparent wind, will give a much higher load on the main of a catamaran than on that of a dinghy. So again, the formula does take the righting moment into account.

Table 7.1 Rod and wire compared at a load of 25% of the breaking load

material	diameter in mm	breaking load in N	stretch in mm per m	weight in kg per 100 metre length
19 strand spiral wire in 316 steel	11	92704.5	2.1	45.1
Dyform 19 strand spiral wire in 316 steel	10	95843.7	2.1	46.6
Nitronic 50 rod	9.5	95647.5	1.74	56.0
MP35N cobalt rod	8.71	98541.45	1.73	50.0

But let us get back to our original subject and check that the shrouds are suited to the dimensions of our chainplates. A 15 mm hole in the chainplate will need wire of at least 19 strands with a diameter of 10 mm, or a rod of at least 7.5 mm. In other words, to find the right pin for 19 strand wire you multiply its diameter by a factor[5] of 1.5, and for rod you multiply by a factor of 2.

But, since we have already hinted at the matter, which is better, wire or rod? Good question! But let us be clear right off: it is a very unequal contest. There is a ratio of 1 : 4 in price between shrouds in wire and shrouds in rod: rod costs from three to four times as much as spiral. When is it justifiable to have shrouds in rod, given the exorbitant cost? Certainly on a racing yacht, where the very low stretch of rod is essential for keeping the mast set up as the crew wants, and certainly also on prestige cruisers where speed performance is obligatory given the class of these yachts. In fact Nautor was one of the first yards to use shrouds in rod although, rather curiously, they were mixed with shrouds in wire.[6]

So let us look at the pros and cons of the two systems. Rod offers two basic advantages: first, it stretches very little under load,[7] and this obviously helps a lot when you need to keep the mast straight, while wire shrouds would not be able to keep the mast as steady. The second advantage is long life.[8] That is right, the very factor that was much criticised when rod made its debut has now become its strong point. Today we can see rigs in rod that have 20 years of life behind them and are still going strong.

The fact is that the material from which rod shrouds are made – Nitronic 50[9] steel – is a steel that has a very high resistance to corrosion in a marine environment, and the way the rod[10] is constructed (as a single piece with no

[5] This assuming that the pins of the rigging screws, or of the terminals of the wire or rod, are in 316 steel.

[6] Very old Swans still have some shrouds in rod and others in wire.

[7] Compared with wire, rod stretches by about one third less at 50% of breaking load.

[8] Others insist on the better aerodynamics of rod, seeing that for a given external diameter rod is stronger than 19 strand wire rope, both because it contains more material, since there are none of the spaces between strands that wire has, and these spaces obviously reduce the amount of material present in wire, and also because Nitronic 50 steel is decidedly stronger that 316 steel. Some insist on the lightness of rod compared with 19 strand wire for a given breaking load: in our opinion, the key aspects of rod rigging are the reduced stretch – in the racing field – and longer life – for the cruising sector.

[9] The Nitronic family of nitrogen-strengthened austenitic steels with Ni-Mn-N (Nitronic 30, 40, 50 and 60) produced by Armco has high mechanical resistance, is particular resistant to wear, has low magnetic permeability and better corrosion resistance. Nitronic 50 (alias 22-13-5) in particular is a nickel-chrome-molybdenum steel commonly used for good brands of rod such as Navtec or Riggarna, while for (extreme) racing use, when rating rules permit, cobalt rod (MP-35-N with nickel, cobalt, chrome and molybdenum) is used.

[10] For many sailors, rod seems a very recent innovation: in reality, back in the 1930s the Nicholson J-class *Velsheda* had rod rigging!

grooves, hollows or crannies, unlike wire, which is by nature a tangle of 'openings' and grooves) greatly limits the amount of salt water, dirt and metal filings deposited on it, and these are the prime causes of corrosion and successive breakages.

Unfortunately over the past ten years we have become used to seeing disgusting rod on boats, with an opaque surface due to poor finishing. In fact, some conscientious riggers polish it by hand before delivering it. But in the beginning, rod shrouds were delivered with a mirror finish, not just for aesthetic reasons but above all because the more polished a material is the less it is likely to suffer the corrosive effects of the marine environment.

It is no accident that before the advent of wire rope produced in the Far East, whose only 'quality' was its low price, the wire made by Ormiston, a noble English company, was (and is still today) so highly polished it was known as 'polished' wire, and had an incredibly long life (as well as an equally incredible price).

More recently, an exotic fibre we have already spoken of in other contexts made its entry into the shrouds sector. PBO rigging has been in use for nearly a decade now at *America's Cup* level, and advances in technology and production techniques are now making it affordable at lower levels. Though still expensive, it offers incredible performance gains, and mega yachts and top racing boats cannot afford to be without it. It is only one quarter the weight of rod or wire rigging of equivalent stretch, and offers many other mechanical advantages. The New Zealand mast builder Southern Spars has set itself apart from the competition by producing shrouds in carbon fibre.

Speaking of companies, in the rod sector just two companies have dominated for 20 years now: Navtec and Riggarna. The former, an American firm, dominated until the first half of the 1980s, but then saw almost half of its market go to the latter, originally a Swedish company that spread into England and remained in the market through its English reincarnation, Ocean Yacht Systems, which uses the Riggarna brand for its line of rod. Both companies use rod made in Nitronic 50.

An alternative to these two giants has recently arrived from a Finnish company, OS, which was not content with developing its own range of terminals for rod but also perfected cold heading presses[11], as well as producing rod in Nitronic 50. Speaking of which we must remember that rod, despite its intrinsic

[11] Strange but true, cold heading presses were rarely sold to riggers but made available to them in exchange for creating a sizeable warehouse store of rod. Obviously clones of the original rod pressing machines are available on the market, but it is better to steer well clear of them. The precision of the dies in rod presses is a key factor in the correct forming of the housing for the splice, so it is better not to run risks with imprecise dies. Currently rod up to –76 is available in tight coils (rod can be coiled in larger coils, with a diameter of 200 times its own, so that once uncoiled it straightens out) and these are then straightened out by riggers with special machines. Above –76 it is only available straight, so to prepare long pieces of rigging such as forestays or backstays it has to be joined with special rod-rod terminals.

mechanical superiority, at the start presented a big challenge to riggers, who had to find a suitable way of attaching it to the terminals.

At first, South Coast Rod Rigging, an English company based in Emsworth,[12] tried to solve the problem by threading the ends of the rod so as to screw it into the terminal. In addition, a right-handed thread was used at one end and a left-handed thread at the other (both were screwed into terminals), so the rod itself could act like the central part of a turnbuckle and tension or ease the shroud as it was rotated. Obviously the range of adjustment was limited, and also awkward to implement, because – to take the example of a forestay – you needed one man at the masthead and another at the bow, and they had to turn the rod in unison with a spanner that fitted into a short squared section! All this business served to save the weight of the central part of a turnbuckle: then they come and tell you that only today's racing boats are set up to kill.

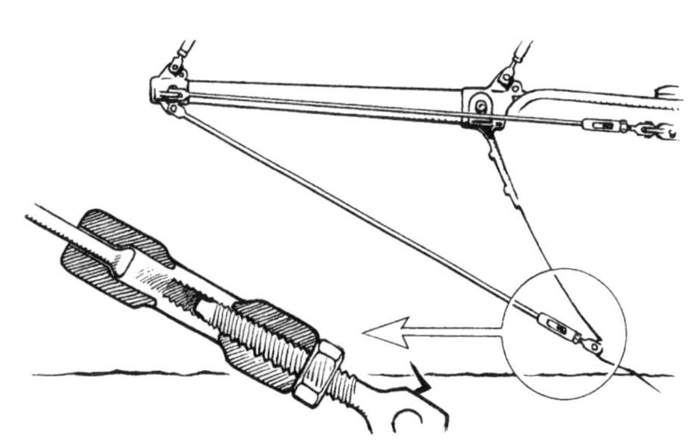

7.2
Section drawing of a rod rigging screw (note the mushroom-shaped head pressed on the rod that fits in the housing in the upper part of the rigging screw).

South Coast also produced profiled rod rigging with an aerofoil section[13] so as improve the aerodynamic performance of the shrouds, which again had the terminal system described above. Lenticular rod was used a lot on the America's Cup International 12 m boats.

However, threading the rod to fit the terminals was not conducive to long life because the threading reduced – albeit not by much – the useful section of the material at a point – where it exited from the terminal – where the fatigue stresses caused by bending as the load direction changed, present even when articulated fittings were used, were concentrated.

It was Navtec that introduced an alternative system to threading. It was called the cold-head rod procedure. And it would be a simple system to implement, if not

[12] P. Johnson, *Ocean Racing and Offshore Yachts*, Nautical Publishing Co., UK, 1972.
[13] This too was in any case a pseudo innovation: at the end of the 1930s the International 12 m Tomahawk already had aerofoil section rod rigging.

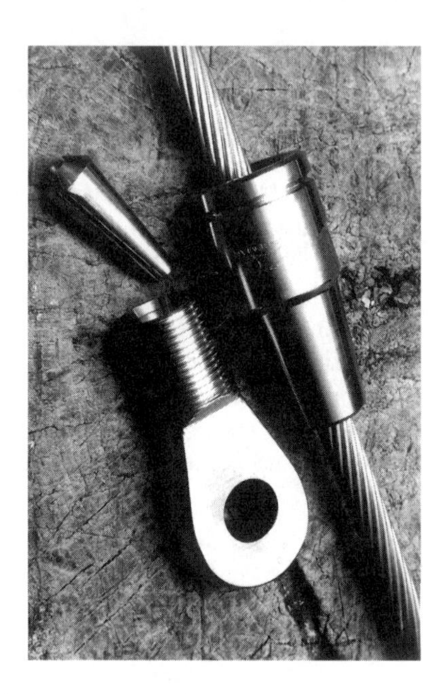

7.3
Norseman eye terminal.

for the enormous pressure needed to stop the rod slipping out of the matrix or template that was supposed to hold it still during the operation.[14]

In practice the rod is held still laterally by a hydraulically powered vice, while another hydraulic cylinder extends to put pressure on the part of the rod protruding from the matrix[14] so as to cold press the material into the housing provided, thus giving it the typical cold head shape.

This cold head on the rod, because it is cold pressed, fully maintains the properties of the original rod, and also has the big advantage of fitting well into the terminal needed, be it a turnbuckle, an adapter for a roller reefing system or one of the many kinds of shroud terminals. At this point it is a simple job: the cold head has to fit into a seat of the same dimensions in the terminal. The rod is thus tightly positioned in the terminal and cannot come out again.

Once pressed, the rod must have a head large enough not to slip through the relevant hole in a special template gauge, which has a series of holes marked for different rod sizes. This ensures the head is of a sufficient diameter. Alternatively, you can simply measure the maximum external diameter of the head with a gauge: it must be one and a half times the original rod diameter.

The axial shape of the cold head is equally important: for example, it must be as far as possible symmetrical with the longitudinal axis of the rod. However,

[14] Each different diameter of rod needs its own pair of dies.

contrary to what you might think, any longitudinal cracks that appear along the cold head (caused by excess material present during the process) are no cause for worry, as long as an expert rigger examines them.[15]

In any case, the cold pressing system, although the best available, did have, especially in the 1980s, its proud – and foolish – opponents. The reason for this lay partly in the high cost of the press needed to work on the rod, and partly in the sloppiness of those who thought they would solve the problem of fitting the rod to the terminal with low-cost home made solutions.

What we are talking about is the disgraceful habit of fitting on to rod swage studs designed for wire. The rod was scored to make it rougher, so that this roughness would guarantee better grip inside the crimp terminal once it was pressed. The tool usually used for crimping wire was then used to press[16] the terminal.

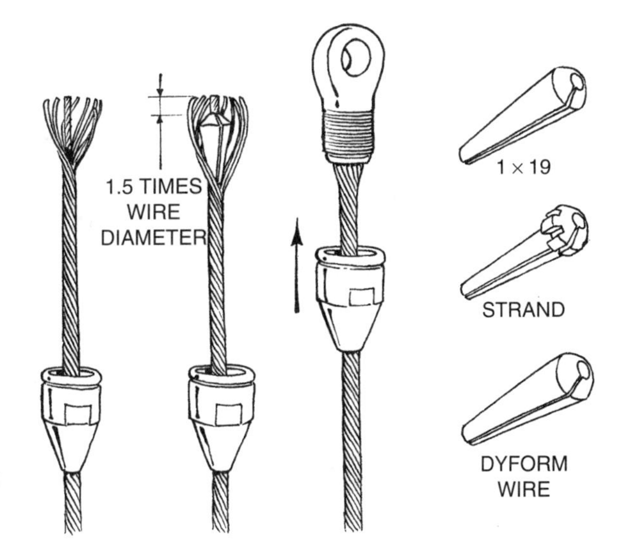

7.4

Fitting a Norseman terminal and (right) various cones for different types of wire.

The debate between the supporters of the two systems was heated, though in fact it never got to the real point at issue: the truth was that a rod with a terminal designed for wire fitted to it had its breaking load reduced to the value of that of a wire terminal. An absolute nonsense. The fact that even mass production yards

[15] The famous skipper with many round the world voyages to his credit, Pierre Fehlmann had the rod rigging of his maxi yacht UBS replaced when he noticed these cracks in the heads of the rod: Navtec hastily denied, in the columns of *Seahorse*, that there had been any risk involved.

[16] Some people, afraid that the rod would slide out, roughened its surface with cuts or abrasions; some put carborundum (basically filings used for the surfaces of grinding wheels, which is actually recommended for use in wire splices). These 'magical' methods say a lot about the lack of seriousness of the riggers who practised them.

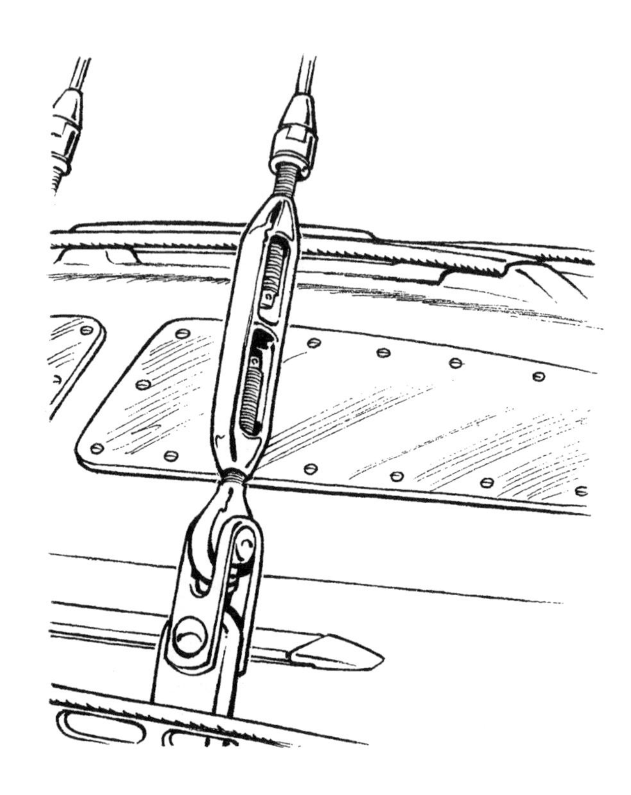

7.5
Rod rigging screw using a Norseman terminal.

gradually abandoned this method of terminating rod shrouds says a lot about the low value, in the long term, of such solutions.

As in many other areas, here too there is a third possibility, half way between cold pressing and using wire terminals. This is the Norseman or Sta-Lock terminal. Both companies produced a terminal for wire that had the advantage of being reusable, since it was fixed on by a completely manual process that did not involve crimping. This means that the Sta-Lock or Norseman could be used again later. The system is in two parts:[17] a female and a male part, both threaded, that have to be screwed together after the wire has been inserted into one of them. First, a conical wedge[18] is inserted into the end of the wire between the core and the outer strands.

[17] This is the difference between the two systems: with the Norseman, the female part contains the cone, with the Sta-Lock it is the male part. Also, the *Sta-Lock* has a cap to place over the end of the strands before closing the terminal; this is very useful for making sure everything goes right when you tighten up the terminal.
[18] Every type of wire needs exactly the right kind of cone: thus there is a cone for 19 strand, for 49 strand, for 133 wire and one for dyform wire.

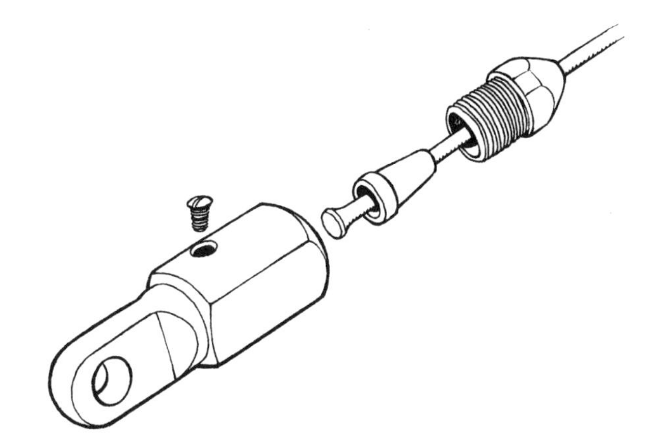

7.6
Sta-Lock rod terminal.

In fact, you have to open the wire with your hands or with pliers, rotating it against the lay, once you have threaded it into the first part of the terminal (which can be the male or female part) so that you can slide the cone on to the core allowing a certain length, which varies according to whether you are using Norseman or Sta-Lock, to protrude. Then you close up the strands again, taking care not to allow them to overlap each other. Above all you must be careful that none of the strands ends up in the longitudinal slot along the cone, which allows it to deform without breaking. Once you have done this you slide the other part on and start screwing them together with two spanners, or hold one part in a vice and screw on the other with the spanner as far as it will go, tightening it down well but not over-tightening; just use your hands on the spanner, not an extension tube.

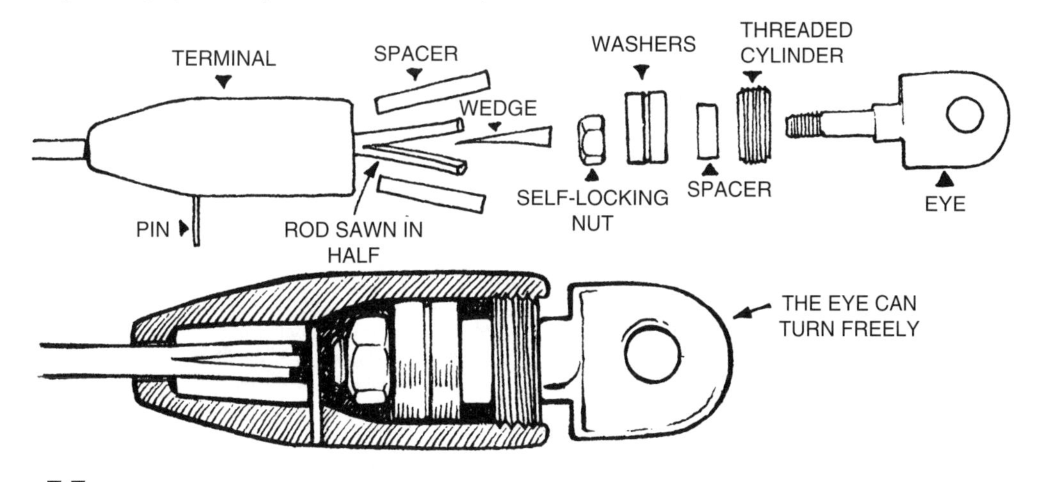

7.7
Twinstay rod terminal.

The thread on the parts is well enough to ensure everything holds together perfectly, so you do not need to tighten it to death.[19] At this point you unscrew everything and do a visual check on the wire inside the terminal to ensure that all the strands are spaced out regularly around their housing and that none of them has ended up in the slot on the cone.

7.8
Wire Teknik hydraulic swaging for wire terminals.

The core needs to protrude as much as is recommended, without any of the strands bending back over themselves. If all these conditions are satisfied, you can fill the terminal will non-acetic silicon (otherwise it will rust inside), screw it up again, this time definitively with a few drops of red *Loctite*[20] on the thread to block it. And the job is done. If the terminal has to be re-used, you simply unscrew the two parts, throw away the old cone and use a new one in its place before screwing up the terminal on the new wire.

Recently these terminals have also made an appearance in the rod sector, where Navtec proposes them – I am referring just to the Norseman[21] – also for cold-pressed rod heads, so as to reduce the cost of rod rigging. As a result Sta-Lock too has adapted its version for rod.

Speaking of prices, beware of imitation Norseman terminals, because though their price is a fifth of that of the original, their quality is just about zero! They are

[19] Here we must mention the Achilles' heel of these terminals: the fact of having a steel thread that fits into another steel thread. You have to be very careful because unfortunately it is not uncommon for these threads to seize: personally I took the precaution of using some anti-seizing liquid when pre-assembling these terminals. Once I was sure it was going to work, I carefully eliminated the liquid before the final tightening.

[20] Loctite, producer of the thread-lock adhesive that bears its name, has in its catalogue a variety of strengths of this adhesive coloured according to their strength, plus other adhesives, including a useful liquid one that glues mechanical bushes.

[21] Norseman, Gibb and Navtec were merged some years ago into a single commercial entity.

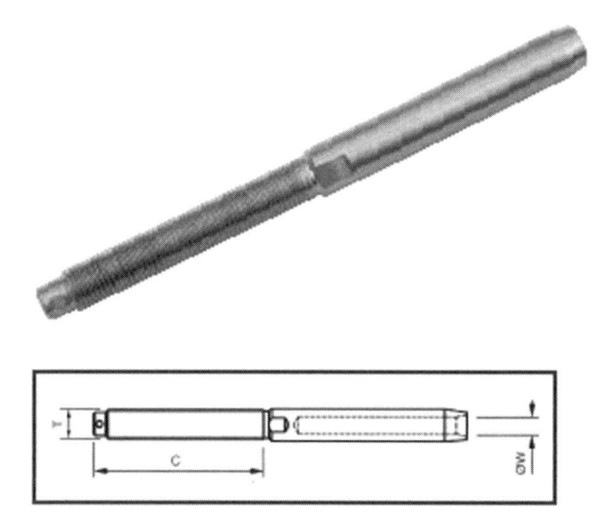

7.9
Pressure terminal for wire shrouds.

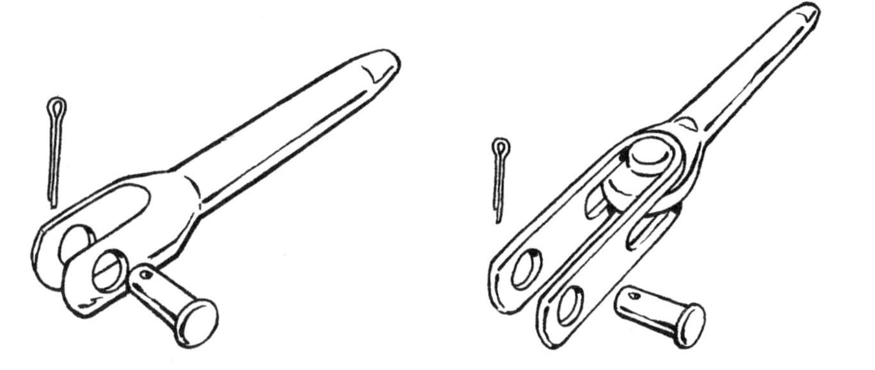

7.10
The difference between a fixed and a toggled fork.

not only much harder to close than the originals, their resistance to atmospheric agents is terribly low.

I cannot end this section on rod terminals without mentioning that rigging iconoclast who goes by the name of Tim Stearn. Besides his very famous masts with three spreaders, Tim designed the Twinstay,[22] a hollow forestay with a groove on each side. The boltrope of the new foresail was slid into the channel facing

[22] The Baltic *Locura* has this kind of forestay still today.

7.11
Rod rigging screw with toggle and a toggle between the chainplate and the rigging screw (superfluous unless it is needed simply as an extension).

forward and, once it was hoisted and ready, the old sail was handed and, as the sheet of the new one was trimmed, the forestay rotated so that the new sail's groove faced astern. The forestay ended with a terminal on ball bearings to allow it to rotate. But the really bizarre thing, as well as the innovation, was that the rod inside the hollow stay was spliced on to its terminal by sawing it in half lengthways, to the depth of a wedge that had to be inserted. It is rather like the Norseman concept, but applied to the rod itself. But Tim produced an even more radical version of this kind of stay: in 1983, on the *Pinta*, in England, he fitted a stay with twin grooves[23] made in just two parts[24] that were glued together with epoxy resin and a special connector. He had come up with this so as to eliminate the internal rod, since the twin-grooved aluminium foil could bear weight and thus carry the load on the stay!

We may ask ourselves at this point, which is better for wire rope, the Norseman–Sta-Lock type or normal swage studs? The former have the advantage of being reusable, though their price is four or five times higher, and greater

[23] On this occasion there were alongside each other.
[24] Stearn's competitor at the time was *Hood*, with its Gemini hollow stays made of bars about 2 m long. This made the system rather delicate, because burrs and deformations at the joins could make it very hard to hoist the sail.

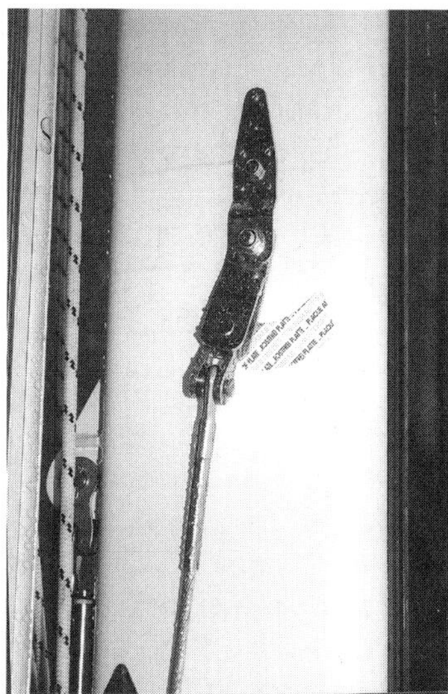

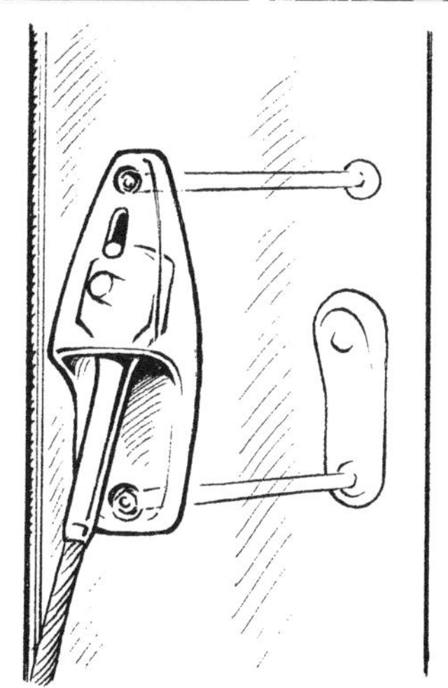

7.12

Toggled fork for rod at the masthead; alongside a traditional mast chainplate for a runner; below: an evolution of the mast chainplate with a plate that houses the eye terminal.

reliability in the long run, since the weak point in these terminals – the point where strands tend to break – is where the wire leaves the terminal,[25] and this does not suffer pressing or crimping as it would with swage studs.

Deciding which is the better system is not so much difficult as simply impossible. Certainly a boat that will be sailed in areas where there are no riggers around will use Norseman or Sta-Lock terminals on her shrouds, and carry on board replacement cones and lengths of wire rope, thus ensuring considerable autonomy and a good safety margin. But the costs in this case would be undoubtedly higher.

At the start of this book I mentioned that marine rigging should never be considered totally separate from the much wider field of industrial rigging. And the origins of the *Norseman* can be traced to the socket, a conical terminal in stainless steel, generally with an eye, into which wire was threaded, unwound, and then molten metal was poured in. Once this cooled and solidified it acted as a permanent stopper. These *sockets* can often be seen in numerous tensile structures and also, though more rarely, on the shrouds of low budget yachts.

The idea of blocking the wire with terminals of this kind, without using a cone but by pouring in a liquid, was taken up again in the 1980s by Loos with the Cast-Lock, into which an epoxy glue was poured. But the lack of a certifiable and reproducible standard for the method of unwinding the wire, together with the fact that the glue deteriorated during storage in retailers' warehouses, led to an early demise for this kind of terminal.

Another variation on the theme of terminals that could be screw tightened without using a special machine was certainly the rapid fix terminal for which the Italian company Douglas Marine obtained a certificate from Lloyd's of London. The wire was threaded into the usual threaded cup, but this time it was held fast by the two valves of a cone whose internal part was well roughened so as to grip the wire firmly, and held in position by a spring. My own specific experience on the forestay of a masthead rig 30 footer was not satisfactory. More recently, the Danish Blue Wave has produced a similar terminal.

The Norseman[26] and Sta-Lock can be found in the same configurations as other terminals. That is: threaded rod (generally in inches but also in some metric sizes), to use for turnbuckles, roller reefing systems or lifelines; male eye terminals, for use with the fork of a turnbuckle or a female chainplate; forked, a female terminal to use with a male chainplate (the fork may be either fixed or toggled); 'T' shaped, with a kind of hook that allows the terminal to be inserted in a female

[25] The end of the splice is a point – both in wire and fibre – where breakages occur both in standing and running rigging.

[26] Norseman – over the years – has become famous for a double Norseman terminal with a central insulating body that serves as an insulator. It is used in pairs on a wire shroud so as to use the length of shroud between the two insulators as an SSB radio antenna.

chainplate fitted into an opening in the mast; stemball, a chalice-shaped terminal that fits into a cup on the wall of the mast.[27]

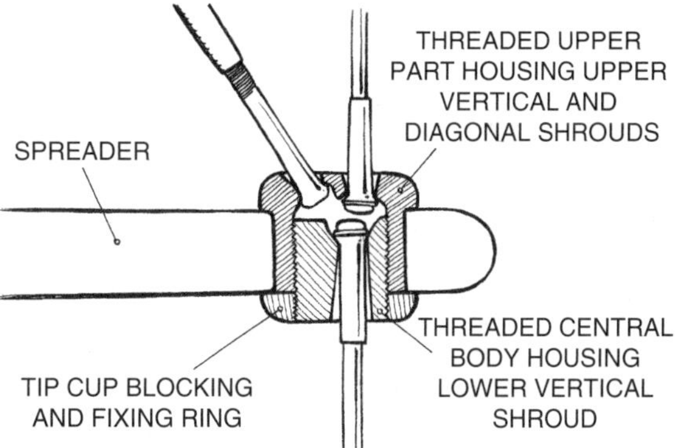

7.13

Tip-cup: joints that allow the shrouds to be interrupted at the ends of the spreaders. Below: section drawing of a tip-cup.

It is easy to understand when to use eye terminals, but there can sometimes be some doubt over whether to use a fixed or toggled fork. I will never tire of pointing out the enormous advantages of a toggled fork over a fixed fork in terms of the life of the shrouds in return for a modest increase in costs, or the addition of a toggle between the terminal of a shroud and its chainplate.

[27] In fact there are other terminal shapes, such as the 'T', but they are rare and little used. They are mainly found in industrial or tensile structure rigging, both of which fall outside the scope of this book.

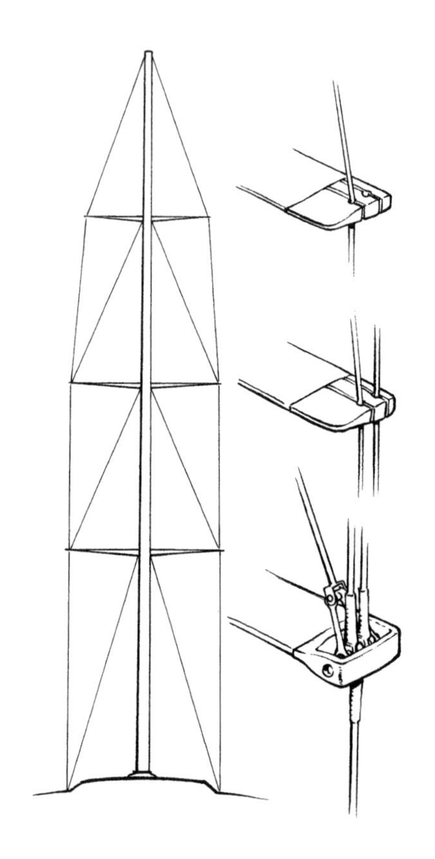

7.14

A mixed type of standing rigging: it is discontinuous at the first spreader and continuous at the second and third.

It is precisely because the weak point of every part of standing rigging lies in the point where the wire (or rod) leaves the terminal, this because of the cyclical[28] bending to which the area is subject during the life of the shrouds, that it is absolutely necessary to have a toggle in there to limit these damaging effects.

In the case of a fork, and in the case of the shroud of a cruising mast, we must take care that the chainplate does not, thanks to the flexibility induced by its length, act as a toggle. Not only would this not avoid fatigue stress on the wire, where it exited from the terminal, it would seriously prejudice the life of the chainplate itself.

In the case of stays, whenever they are designed to have sails hanked onto them (which would mean sag caused by the sideways pressure of the sail) it is

[28] The importance of the concept of fatigue, or rather of cycles of fatigue, is easy to understand with the example of a strong metal wire: if it is bent once or twice it does not break, but if you bend it repeatedly several times it will suddenly snap.

absolutely obligatory to have a toggle at the base. And I mean 'at the base': the toggle must be inserted between the chainplate on the deck and the wire, not aloft as I have often seen done. The toggle is needed at the bottom, not high up! What is true on the deck is also true on the mast, so a toggle will also be needed where the stay joins the mast.[29]

7.15
Steel spreader bend on tip of spreader.

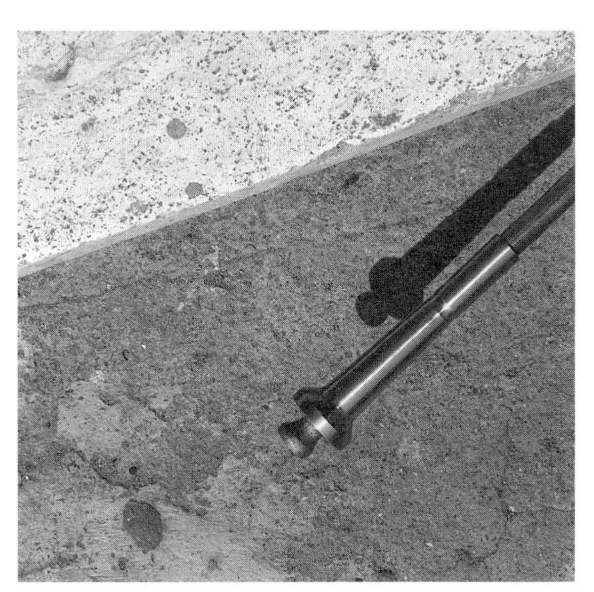

7.16
Rod head inserted in a stemball sleeve.

[29] Some applications apparently do without the toggle since the stay terminals are fitted to special pins with the central part barrel-shaped: these ensure a certain degree of pivoting with the advantage of being much less bulky than a toggle.

There is not much to say about 'T' terminals: they are simply the most economical solution that a mast builder or a rigger has for rigging a stayed mast. The mast builder prepares a simple circular hole in the mast to accept the shroud terminated with a 'T' terminal, mounting a special plate inside this hole. If used for a sidestay, the terminals must not have to work beyond plus or minus 20° with respect to vertical in the fore-aft direction and between 5° and 25° laterally: this is their critical point. This is in the sense that, if they have to work outside this range of angles, it can cause cracks in the terminals themselves or shorten the life of the stays. This is why it is never advisable to use a 'T' terminal for an inner forestay or a babystay, though unfortunately this is often done.

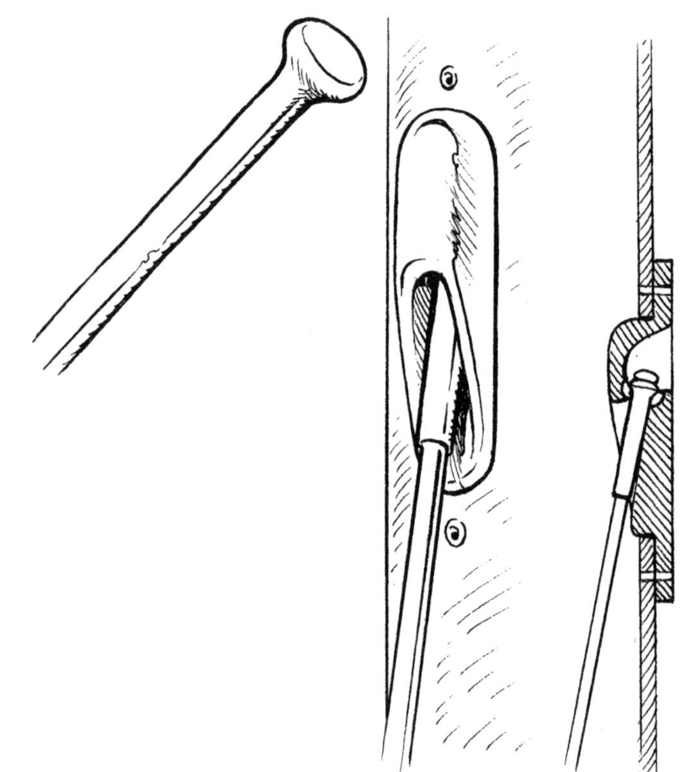

7.17
Rod ball head: a special racing head without stemball; right: lateral rod rigging attachment for cruising boats (basically an oblong shell with a housing for the stemball).

More or less the same considerations apply to stemball terminals. The stemballs, or rather the cups that house them, are unlikely to permit a sufficient degree of play for the sidestay, and if you take a turn around the quayside you will see many sidestays on boats with these terminals that bend and press tremendously on the point where the wire leaves the terminal: and that means a short life for the wire!

7.18
Shroud attachment with rod head in stemball.

But speaking of the life cycle of shrouds brings us back to rod: how can you check what state your rod rigging is in?

Knowing what kind of boat we are dealing with and what use has been made of her is already a help. But the best thing imaginable is to unstep her mast, and thus dismantle all the shrouds, and look at the heads of all the rods once they have been removed from the housings in their respective terminals.[30] We can take an even closer look by using penetrating liquids: there are three or four of these, according to the brand, to use in succession. After a final rinse with fresh water they will reveal any micro-cracks present, though it is not always easy to attribute these to the beginnings of real cracks since sometimes the cold pressing of the rod leaves a surplus of material on its surface, a kind of lip or crest, and this will be shown up by the liquids just like a crack and may deceive the inexpert eye. But one element to bear closely in mind during the inspection is the kind of rod present on the boat: is it continuous or discontinuous?

Continuous shrouds, that is when the various sidestays start from the chainplates on the deck and go right up to the mast, passing through the ends of the spreaders without interruption, offer a simpler and more economical solution than discontinuous ones – where the shrouds are interrupted at every spreader and a new wire starts leading up to the mast – since they do not need tip-cups. Tip-cups are basically joints, more or less complex and sophisticated according to use

[30] The X-rays many people got enthusiastic about at one point have demonstrated their ineffectiveness: they only manage to show the end of the life cycle of the rod when it is in a terminal state and about to break anyway.

and needs, that allow the sidestays to be interrupted at the ends of the spreaders and a new section to set off towards the mast, at the correct angle and thus with a suitable range of toggling to permit correct alignment.

In 1983 Seahorse published an article by Peter Morton, an English rigger who revolutionised the way rigs were understood. He demonstrated, data in hand, the superiority – in the racing field – of discontinuous shrouds over continuous shrouds, hinging his argument on something that everybody in the sector was very sensitive to: stretch. Discontinuous shrouds stretch much less than continuous ones of the same diameter and material. This is for the simple reason that each side shroud – starting from the D2[31] – works with a point to point length – that is, the distance from the end of the spreader to the side of the mast next to the upper spreaders – that is far less than that of the same shroud in continuous form.

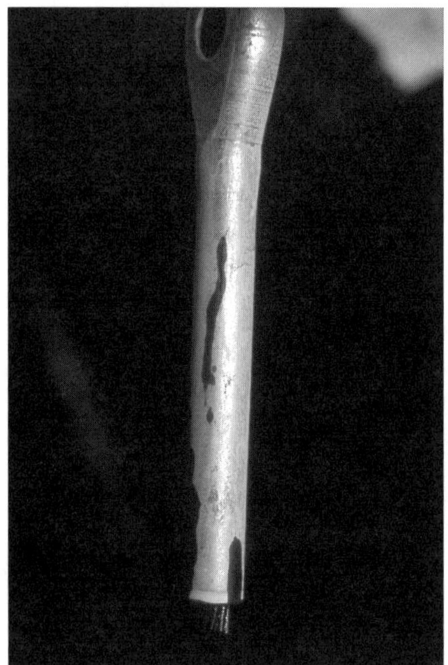

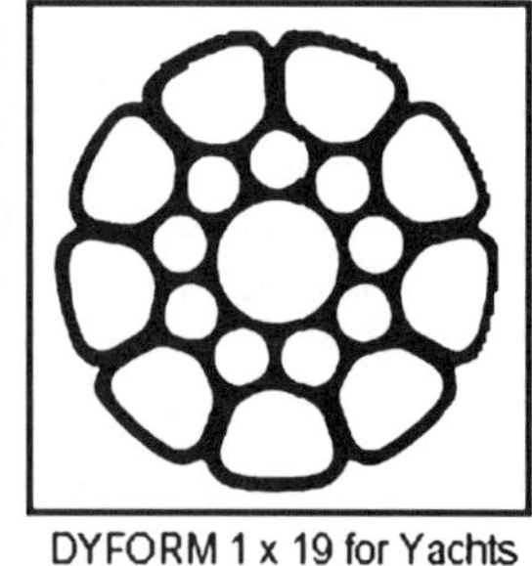

DYFORM 1 x 19 for Yachts

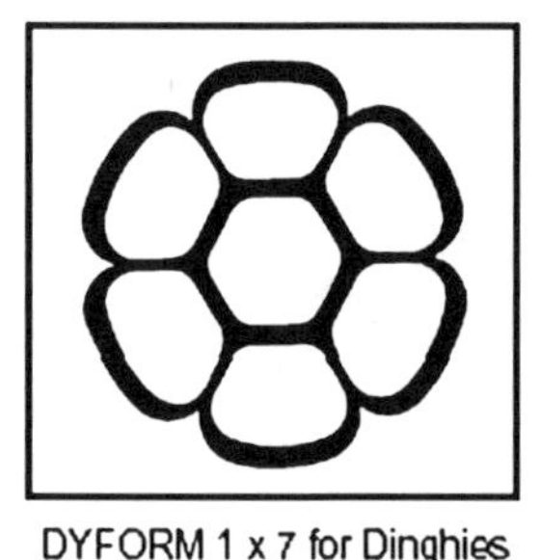

DYFORM 1 x 7 for Dinghies

7.19

Wire terminal tested with penetrating liquids; right: 1 × 19 and 1 × 7 dyform wire.

[31] Discontinuous shrouds are denominated D1, D2, D3 and so on – these are respectively the low, lower intermediate and upper intermediate diagonal shrouds. V1 is the lowest part of the vertical shroud that goes from the chainplate to the first spreader; V2 the part that goes from the first to the second spreader and V3–D4 are the final part of the vertical shroud, between the second and the third spreader, and the diagonal shroud that is fixed to the mast.

The big drawback of continuous shrouds is not just greater stretch; it also lies in the fact that, to pass over the ends of the spreaders, they need a *spreader-bend*.

The *spreader-bend* is a specially curved tube that serves to distribute the forces – which would otherwise be exerted on too small an area of the rod – over a larger surface, thus diminishing their intensity and consequent negative effects. But in the long run the suspicion remains that the part of the rod that follows this curve is prone to cracks.

The use of discontinuous shrouds and tip-cups means less stretch and greater reliability than with the continuous type, but with greater weight, cost and a higher centre of gravity, together with not indifferent problems of regulation, since every time we want to regulate a diagonal shroud, that is not the D1 or the V1, we have to shin up the mast.

In fact, many producers are nowadays offering a return to continuous standing rig, as seen typically with rod in the 1980s, eliminating tip-cup fittings.

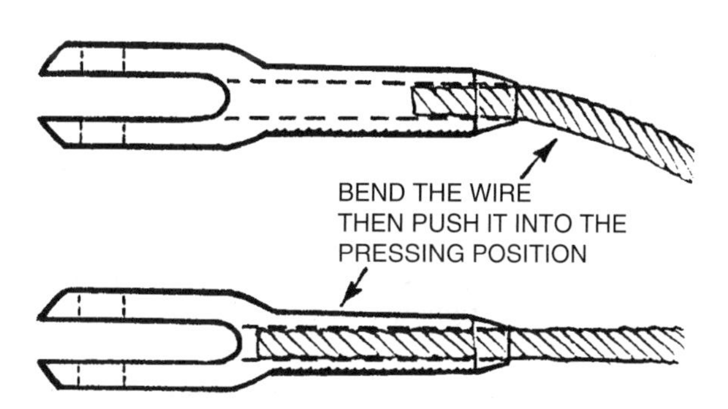

7.20

It is important to insert the wire correctly, bending it slightly when it is partly inserted so that once it is fully in, it will not slip out of the terminal while being pressed.

Speaking of precautions to be taken to ensure the long life of rod, it is well to remember the use of the *stemball*. This has the form of a truncated cone, and houses the head of the rod. It considerably prolongs the active life of the rod, thanks to a better distribution of the forces per square millimetre due to the greater surface of the stemball.

A very interesting version of the stemball – which few people know about – for cruisers is the so-called *high-fatigue stemball* which has a longer body than the stemball used on racing boats and thus is heavier. Its central part has a cavity specially hollowed out to provide a 'fuse' zone. The fuse zone is specially made more susceptible to cracks without interfering at all with the overall strength of the terminal. This means that at the first sign of fatigue, the appearance of a crack there warns us of the danger well before we run into decidedly more deleterious consequences.

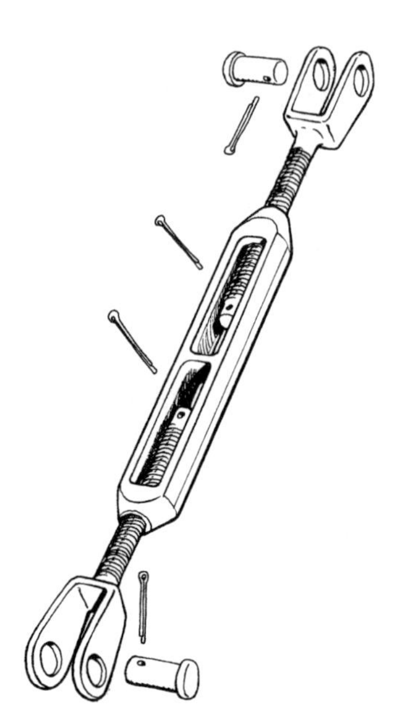

7.21
Hasselfors fork-fork terminal.

At the opposite extreme – purely in the racing sector – we find the stemball completely abandoned and replaced by a hemispherical head pressed on the rod with special matrices to form a shape that goes some way to make up for the lack of a stemball by having a larger surface than the traditional mushroom-shaped head. But it is reserved for use on boats that will only sail a few offshore races.

Now we have finished our look at rod rigging, what can we say about wire shrouds other than to describe the strong points that have made them so popular? I am talking about the substantial cost savings compared with rod; then there is the reliability factor to take into account, which derives from the very way the wire is made with its several strands. This means wire is unlikely to break suddenly as rod does, and the strands actually show the ageing or the poor quality of the wire by breaking in a limited number that usually means the wire can be replaced before any irreparable damage is done.

In other words, wire has – so to speak – an inbuilt 'fuse' that will signal the end of its life cycle: one or more of the strands most subjected to stress.

Today it is common practice to consider as a synonym for shroud wire, 19-strand stainless steel wire rope. But in fact this is only a convention, and it is not uncommon to find traditional boats rigged with 133 strand galvanised wire, which

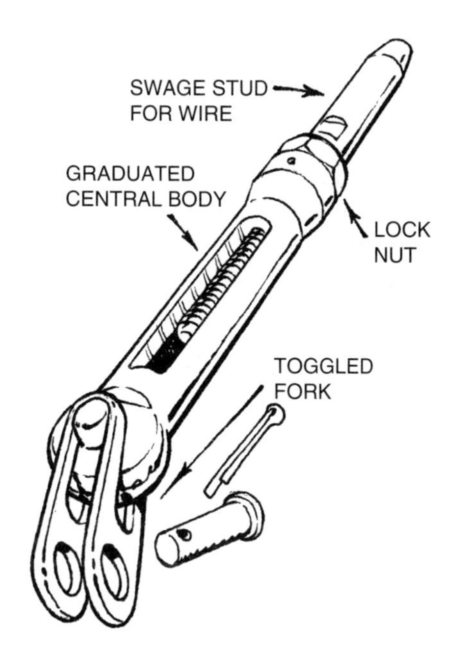

7.22
Ronstan Sealoc rigging screw with adjustment markings.

is usually hand spliced. If a world-voyaging owner knows how to splice by hand and knows something about the maintenance of galvanised wire, he can, at very little cost, be completely self-sufficient in whatever part of the world he sails. And we should not be surprised to find 37-strand wire, especially on sailing ships, since above a certain size rope makers produce this kind of wire in place of the better know 19-strand version. Still on the subject of the apparantly bizarre, we may find standing rigging in 49-strand wire for the simple reason that this is also very common in non-marine sectors.

We should add a note on the material used for wire ropes. In Europe 316 steel is most used, while in the United States 302 or 304 steel is common. The difference between the two types lies in the greater resistance to corrosion of 316 steel, which is an important factor in Europe where boats that sail the Mediterranean, where the salt content of the water and the temperatures are high, are more under attack from the marine environment. Wire rigging in the Mediterranean – in the quality of wire that is by now standard in Europe – lasts on average not more than six or seven years. In the United States 302 or 304 is generally chosen because of its better mechanical qualities, and its lesser resistance to corrosion is tolerated.

Finally, in wire as in rod there are touchstones in world production. At the top in Europe I would place the small company Ormiston and the giant Bridon[32] – both of them British – and in the USA McWhyte, Loos and Carolina.

A few last words on ways of splicing wire shrouds. Besides the Wire Teknik style swager with rotating dies, riggers who have been in business a long time still use the hammering machine. This looks rather like a lathe. The terminal and wire are inserted in an opening, and a series of hammers in rotation reduce the terminal to the necessary diameter over the required length. The wire does not slip out of the terminal because metal is smeared over the irregularities in the wire so as to form a single amalgam between wire and terminal, so there is a friction based grip that will last over time.

7.23
A rod shroud attached to the side of the mast.

Other systems used are based on industrial presses that compress dies with a hexagonal section and thus reduce the terminal to the section required. This certainly makes shroud production much faster than with a hammering machine,

[32] Bridon is to be credited with the invention of a special compacted wire called dyform, which has the peculiarity of having its external strands with a polygonal and not a circular section. This means that dyform fills the cavities that would remain open in normal wire and, for the same diameter, thus incorporates a greater amount of material and this increases the breaking load and reduces stretch, again for a given diameter. But note that you pay the price of increased weight. It can be useful in strict one design classes like the *J24*, where the maximum shroud diameter is set at 5 mm, so using 5 mm dyform you have less stretch, though it does weigh more. The downside of this option, apart from greater weight and cost is that, like all wire, dyform under tension has its strands oriented in the direction of the load, so they rub against each other, and in this movement the edges of the outer strands, where the genoa passes, can chafe against the sail during tacks and damage it. Because of the greater breaking load of dyform compared with standard 19 strand wire, it is recommended that terminals of top brands be used so as to ensure the breaking load of the terminals is fully compatible with that of the wire.

but the terminals will also age much faster because the sharp edges present on them will cause premature cracks. To perform a job on the terminal of uniformity comparable to that of the hammering machine without its epic time frame, for many years a series of rotating die swagers was used. The terminal was inserted and both guided and pressed by the dies until it reached the required diameter.[33] But this led to a defect: the terminal tended to come out bent like a banana because one die fastened on to and then pressed the terminal before the next die. Quite apart from the horrendous aesthetic result, the more the banana shape put stress on the wire where it left the terminal the more serious was the defect. It shortened the life of the wire, which always broke where the terminal pressed against it.

All this ended when a gentleman in Sweden by the name of Leif Anderson had the brilliant idea of revolutionising the way terminals were pressed simply by producing a swaging machine – the Wire Teknik – that pulled the terminal (with a hydraulic cylinder) through a pair of dies that were fixed together but free of any servomechanism. In this way the terminal left the swager straight as a die. A Copernican revolution.

Turnbuckles or Rigging Screws

'On modern sailing ships, from the mid-nineteenth century, rigging screws were brought in instead of deadeyes'.[34] The rigging screw is basically made up of two threaded rods, one with a left-hand and the other with a right-hand thread. These are screwed into a central body. This central body can be turned by hand to tension the shroud it is fitted to. The rigging screw can have both or just one of its rods threaded, and is available in all the versions described above for terminals. The central body may be of the closed type, so that the threads are protected inside it all along the length that is screwed in, or open, with the threads always visible.

A big step forward in rigging screws came from Riggarna with the CR4 model. This rigging screw did without the upper threaded rod, replacing it with a simple sleeve that housed the rod or wire, while the central body was rotated as usual to tension the shroud, screwing down on to an extra long threaded rod with

[33] Generally optimal pressing reduces the external diameter of the terminal by a minimum of 10%. Riggers have special tables supplied by the makers of splicing machines to check the exact reduction in diameter according to the terminal used. It is also very important that no pressure is applied to that part of the terminal that is not hollowed out to receive the wire: this would produce anomalous movements in the material the terminal is made from and lead to premature cracking. The user must check that the terminal is straight and can also easily check that the diameter of the pressed part is 10% less than that of the unpressed part close to the end of the terminal.

[34] O. Curti, *Il libro completo dell'attrezzatura navale*, Mursia, Milan, p. 121, 1979.

a toggled fork. This rod was threaded only on its extreme upper part, and this – together with the considerable length of the central body – meant the CR4 presented no exposed thread.

Using concepts similar to that of the CR4 companies, such as *Searig* in New Zealand and *Ronstan* in Australia, made rigging screws of a similar kind. It is worth noting that all these kinds of rigging screw had their central bodies in bronze, which meant they could act on the threaded steel rod without risk of seizing, which is common in the case of rigging screws made entirely in steel.

Setting up a swept back rig

Here is How to Set Up the Mast

Here is a description of how to set up a mast, fractional or masthead rigged with aft raked spreaders. This kind of rig is very popular nowadays, because since it needs no runners it fits well into the 'easy sailing' philosophy. I have decided to explain how to set up this rig, which is a rather complicated business, because when it comes to simpler rigs the procedure remains the same and you will be able to set up any other mast and rig with the same basic techniques.

In essence you have to achieve the effect of having the masthead at a point along the centreline of the boat in the athwartwhips sense, and the same alignment for the point of attachment of each set of spreaders. You must also achieve the correct degree of rake, that is, the fore-aft inclination of the mast.

Since there are no runners, the tension of the forestay is felt mainly on the upper shrouds, and because these are oriented aft, they are able to bear the load, helped to a lesser extent by the backstay. But for the privilege of not having runners, there are the following prices to pay:

1. The mast is very hard to set up, because any adjustment of shroud tension inevitably means a change in the curvature of the mast in the fore-aft sense, with all that entails in terms of flattening or powering up the main;
2. There is more sag in the forestay compared with a traditional rig.

8.1
Swept back rig.

Here is How to Set Up the Mast

A. Once the mast is stepped, tighten the rigging screws by hand – after lubricating all the threads with Rig-lube[1] spray to prevent seizing – until the mainsail track looks straight. You should sight along the main track from bottom to top, which is looking from the boom up towards the masthead. In other words, try to set up the mast by eye so that it is along the centreline of the boat in the athwartships sense. The system will need to be refined, since you will now increase shroud tension using the rigging screws, and this demands an effective way of checking that the mast remains along the centreline.

Generally it is far preferable – when possible – to insert a slide into the mainsail track (or free up a car if the main is fully battened), and use the halyard to hoist it up to the masthead after fixing to it a metal measuring tape of suitable

[1] Rig-lube is definitely the best spray for rigging screw threads. It is made by Navtec, and based on molybdenum. It lubricates without draining away.

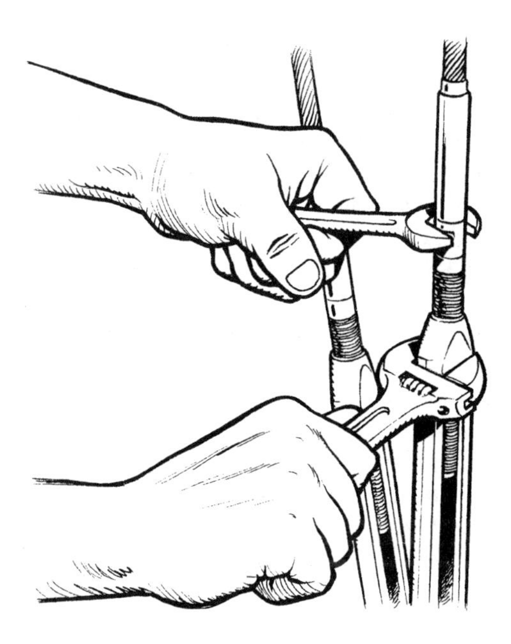

8.2
Tensioning shrouds using two spanners on a rigging screw.

length. It is important to use a slide, because the main halyard often exits from the masthead off to one side; this is typical on boats with masthead sheaves both for the main halyard and for the topping lift. And this means measurements will differ from port to starboard, even if the mast is correctly centred! Using a slide or car in the main track avoids this problem.

In addition, by bringing the slide down from the masthead along the track and stopping it at the level of the various sets of spreaders, you can carry out the checks needed to ensure the mast is centred also at intermediate points.

Unfortunately, it often happens in the case of aft raked spreader rigs that the measuring tape, when stretched from the masthead, ends up pressing against the upper shrouds, and this makes serious measurements impossible. In fact, the whole point of the measuring tape is to measure the distance between the masthead (or the point of attachment of the spreaders) and points along both sides of the boat, which could be the points where the chainplates are fitted or any other points equidistant from the centreline identifiable on both sides of the vessel. If the distances measured to port and starboard are different, the mast is not centred.

An example: if to starboard you measure 25 550 mm between the masthead and the chainplate, and to port 25 450 mm, your masthead is out to port with respect to the centreline, and needs to be corrected to starboard by tensioning the upper shroud. If the upper shrouds are getting in the way of your measuring tape, try using the genoa halyard. If there are two genoa halyards in the centre and two spinnaker halyards to the side, obviously none of them will be on the centreline. In this case the best thing to do is to fix your measuring tape to both genoa halyards and hoist them together; this will ensure the tape is positioned centrally. Use the tape carefully to control the distances to port and starboard, and be sure to choose a day with very little wind for the operation, otherwise the tape will sag too much and the measurements will not be reliable.

B. Once the mast is aligned along the centreline at all points, you can set the pre-bend by inserting the wedges in the step in the fore-aft sense, without forgetting the lateral ones that are needed to keep the mast on the centreline at the level of the step. These wedges must hold the mast still in the step, and so need to be inserted with a certain degree of force: if you have problems doing this, pass a line around the mast and tension it on a winch so as to create the space needed to insert the wedge. Attach the main halyard to the gooseneck and tension it slightly: this will allow you visually to check the bend – by sighting the distance between the halyard and the main track half-way up the mast – to ensure it is in line with what your sailmaker requires. If you do not have any indications from him, it is better not to worry about pre-bend and concentrate on keeping the mast straight, not bent, in the fore-aft sense. Use a tape, or better still a gauge, to measure how much thread is left outside the bodies of the upper shroud rigging screws, both to port and to starboard, and tighten them on both sides of the boat by three-quarters of a turn.

While one person does this job, another must sight up the main groove in the mast as described earlier. Check in particular the increase in mast bend shown by an increase in the distance between the halyard and the groove, and the centring of the mast by measuring with the tape as described earlier.

Stop tensioning the uppers when you reach the tension suggested by the mast builder or, if you do not have this value, tension them until you have taken out the slack you feel when you pull on the shroud and it bows noticeably towards you. Then start tensioning the lower shrouds, always keeping an eye on the mast, and then also tension all the intermediate diagonal shrouds until you have taken the slack out of them but no further. If the rig has more than one set of spreaders there may be, in a traditional rig, an inner forestay that is very useful both for setting the initial pre-bend and for a final tune of the fore-aft bend.

At this point, static tuning at the mooring is complete and you need to get out on the water, ideally on a day when the water is flat and with 10 to12 knots of true wind so you can sail the boat close hauled with full main and genoa and the crew in position. While sailing in this way, sight up the main track to check that the mast is not 'falling off' to leeward at any point: if it is you will need to tension

8.3
Danilo Fabbroni, to port, and Vittorio Vongher to starboard, engaged in setting up the diagonal shrouds of a maxi yacht.

the corresponding shroud, and to do this you will need to go about so as to take the load off that shroud. When you tension a shroud on the leeward side, do it very cautiously: tighten by one or at most two turns at a time, because when you go about again the shroud could be overtensioned and be a danger to the mast.

A mistake that is more common than you might think is over-tensioning the lateral shrouds and thus excessively compressing the mast, which gives it an excessive 'banana' bend in the fore-aft sense. If you notice that the masthead keeps 'falling off' to leeward although the upper shroud is well tensioned, and you keep on tensioning it, stop. All you are doing is bending your boat and stretching your shrouds! Often in these cases the only thing to do is to slacken off all the shrouds and start again from scratch. Another clear sign of excessive compression is when the ends of the lee-side spreaders are further forward than their points of attachment to the mast: at this point the mast is practically turned on itself: the leeward upper shroud is not supporting it because excessive tension in its windward counterpart has left it slack. Again, beware of over-tensioning.

Concerning the backstay, be careful not to tension it beyond the point at which it no longer increases the tension in the forestay: more tension than this would be quite useless and also dangerous. Note in this context that many owners of aft swept rigs insist on trying to remove the sag from the forestay by trying to tension the backstay instead of the forestay itself with hydraulic cylinders or other below-deck systems.

Winches

The archetypes of present-day winches[1] were the windlasses present on sailing ships of past centuries. A vertical cylinder was turned by sailors pushing wooden bars. These bars served both to rotate the drum of the windlass and to prevent it, thanks to the men operating it, from rotating in the opposite direction under the load applied to it by the line being tensioned. In other words, the bars served as the precursors of the pawl, the small lever found in modern winches.

The first winches designed for yachting had two points in which the handle could be inserted: one linked it directly to the drum and gave a 1 : 1 ratio, while the other used an internal gearing mechanism to provide a different ratio. Companies such as Goiot in France and Merriman in the United States made winches of this kind. But on a racing boat, having to stop and change handle position in the middle of a tack was clearly a big disadvantage. Well aware of this fact, in 1959 Tim Moseley, owner of the yacht *Orient*, asked the technical director of his San Francisco mechanics company, Jesus Guangorena, to design a winch that was innovative in this respect. This gentleman had an idea that was definitively to revolutionise winches. He added two freely rotating wheels inside the two gearings inside the winch that were constantly engaged. The freely rotating wheels allowed the gearing in which they were inserted to be actioned only in the opposite direction to that of the other gearing. As a result, turning the handle in one direction gave one ratio, turning it the other gave the other.

Tests of the prototype were so encouraging that Tim Moseley set up in business with a friend to produce these winches. That friend was Jim Michael, owner of the yacht *Baruna*, and thus was born Barient, from a contraction of the names of their two yachts.

When the handle turns clockwise, the pawls of the freely rotating wheel at the top of the winch turn the drum with a 1 : 1 ratio and the pawls of the lower freewheel click in neutral. When the handle is turned anticlockwise, the opposite happens and the drum is turned with a 5 : 1 ratio. What was innovative about this system in mechanical terms was that this combination of the two freewheels meant

[1] The term 'winch' seems to be derived from the Old German *winde*, windlass.

the drum could only rotate clockwise, and thus could turn in this direction either under the action of the handle rotated in either direction or freely, as happens when you trim a sheet by hand at the start of a tack.[2] This system is still in use today in all winches with at least two speeds. In larger winches, where a ratio different from 1 : 1 is desirable in both speeds, a more evolved system is used. Here too, the pair of freewheels automatically changes the ratio, prevents the drum from rotating anticlockwise and permits it to rotate clockwise.

9.1
A pawl with its spring.[3]

The next advance in winches came with a device that allowed direct drive between handle and drum to be selected. This meant a manoeuvre could begin with a 1 : 1 ratio, and when the handle was then turned in the opposite direction it reverted to the original scheme, thus producing a three-speed winch. In practice, if you start turning the handle without having pushed the button[4] on the upper part of the winch, the winch will always be in second or third speed, depending on whether you turn the handle clockwise or anticlockwise. If you push the button, a mechanism engages first speed and the winch will operate at first and second speeds. Once you operate the winch at second speed it will no longer offer first speed, thanks to a mechanical release that can be of various kinds. In practice it is best to engage the first speed button before tacking or before a fast haul and start out that way, ending up with third speed for fine tuning.

[2] This is known as free-spinning or free-tailing. It is useful to spin the drum by hand in this way as you can hear whether the drum is free from friction, especially between the pinion gear of the winch and the ring gear of the drum.
[3] The straight edge of the spring must always be placed against the pawl, the curved part going against the part of the gear where the pawl is housed. Pawl springs must never be greased but merely oiled with normal sewing machine or weapon oil.
[4] The third-speed button may well be on the base of the winch.

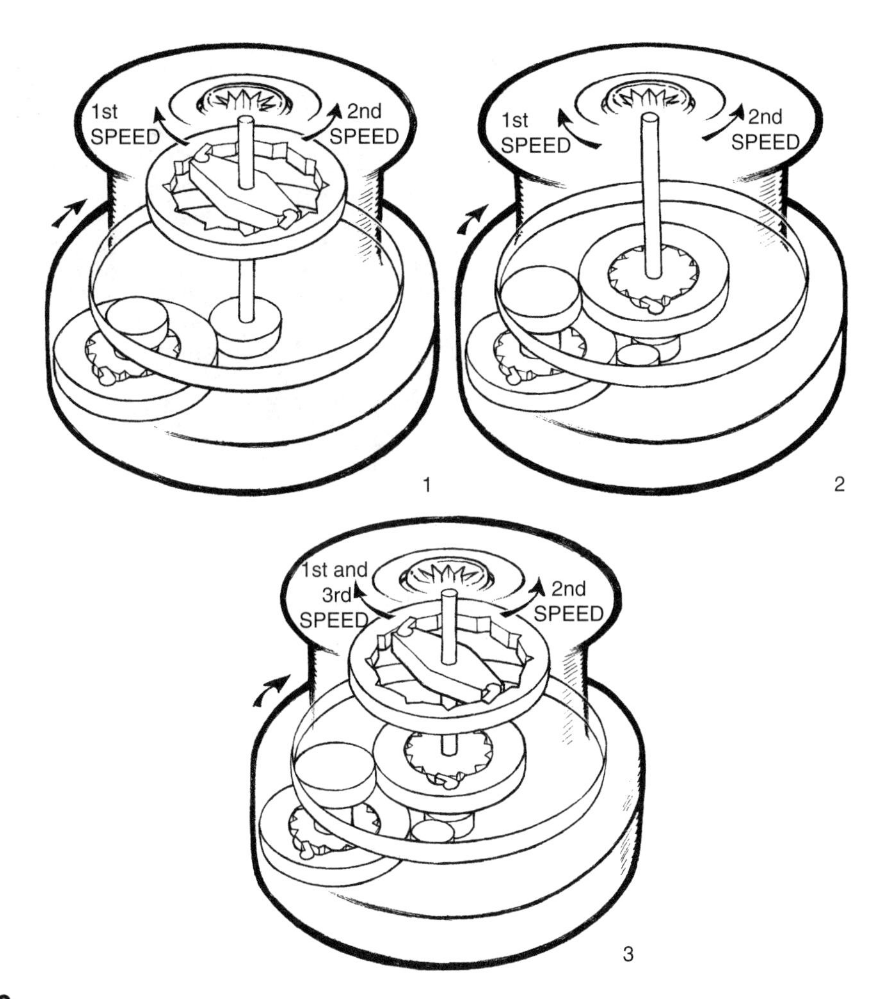

9.2

Section drawing showing the gearing of a winch in the various use modes.

In some cases, a winch that behaves in this way is deleterious. For example, when trimming the chute on a reach: if the winch goes into second and third speed mode, third is too slow for trimming the spinnaker sheet and we would prefer the winch always in second and first speed mode. So along came another gadget called the *lock*. This is a button on the top of the winch, or on the base, or a line that comes out from under the winch and is led to a jam cleat. Once activated it keeps the winch in second and first speed mode. Later winches with

9.3

A three-speed 65 winch; third speed is engaged by pressing the black disc on top of the winch.

four speeds[5] basically use the same system but can select one of the two possible first speeds from the three to be used during the manoeuvre.[6]

The power ratio[7] of a winch – usually shown as a number on the top of the winch[8] – is the mechanical advantage, given by the following formula:

Power ratio = Gear ratio × handle length/drum radius + sheet radius

where the gear ratio[9] is given by:

Number of rotations of handle/Number of rotations of drum

[5] Generally a four-speed winch is a three-speed model with another, reduced first speed added so that the first speed with the power ratio best suited to sailing conditions can be used. Be careful not to confuse the lock control, which locks the winch in first and second speed, with the 'gear shift' control, which switches the winch between close hauled and reaching mode.

[6] B. Ottemann, 'Il misterioso interno di un winch', pages 16–8, *Technical Sailing*, N. 2 1996.

[7] This is not always true because sometimes, for reasons of product numbering, a number is used that does not exactly refer to the power ratio. For example, the *Harken 990* winch has a power ratio of 100.

[8] An exception was the now defunct American company Barient which placed on the winch top a number, 22 for example, which meant that the winch could be used to trim a sheet with a working load of 4400 pounds. This figure was arrived at by multiplying 22 by 100 and then by 2. This was an attempt to give the customer an idea of the final efficiency of the winch, already taking its loss of efficiency into account.

[9] In many cases, this transmission ratio is called 'speed ratio' with an incorrect use of the term 'speed' that leads many sailors into an incorrect evaluation of the winch.

The power ratio is simply the reduction effected by the winch to reduce the load and thus allow the trimmer to tension the line as needed. It is directly comparable to the reduction given by a purchase. A four-part purchase reduces the effort needed to haul 100 to 25 kg, while a winch with a power ratio of 8 reduces the effort needed to haul 100 to 12.5 kg. Obviously this is the theory, because the efficiency of a purchase or of a winch is never equal to 1 and thus there will be a loss of performance. We have already talked about this for purchases, while with a winch the loss of power ratio will be around 20 to 30 per cent, depending on the model.

Thus in reality, with a 20 per cent loss of efficiency, when we apply a force of 12.5 kg to our winch with a power ratio of 8, we are no longer applying the theoretical 100 kg to the sheet, but only 80 kg.

If you look at the illustration below you can see where the mechanical advantage or power ratio of the winch comes from: from the length of the handle compared to the diameter of the winch drum it is being used on! It is just a matter of the advantage given by a simple lever! The lever is the handle, the centre of the winch is the fulcrum, and the power is developed at the outer point of the drum. The longer the handle (lever), the greater the mechanical advantage (though you will be slower turning the winch). This is why there is no sense in talking about the power ratio if not referring to the length of the handle selected for use. Generally the standard handle is 10 inches long, but for certain applications such as halyards an 8-inch handle is fine.[10]

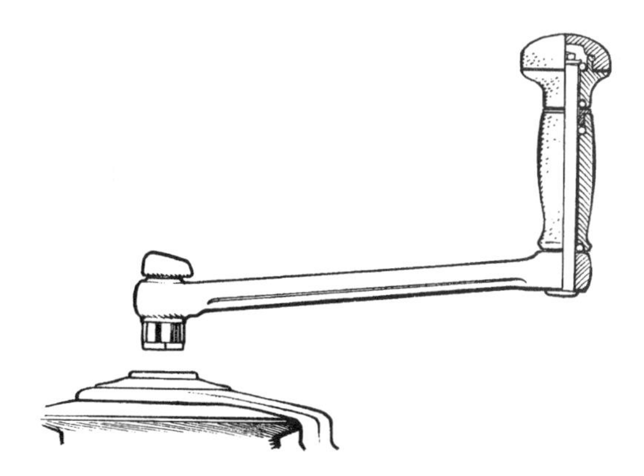

9.4

Detail of a Harken double winch handle (note the speed-grip) that permits a lower position over the winch that improves turning performance).

The handle can be made with its body in chromed bronze, aluminium or carbon. Handles may have a lock at the end consisting of a square plate activated by a lever and a spring, or be without. Handles with locks are slower to insert into the winch but do not get lost, while those without are faster to insert but risk ending up in the water. Handles may be single (with just one grip) or double

[10] One inch equals 25.4 mm.

(designed to be gripped with two hands). There is also a third kind called 'speed grip', an innovation from Harken: the upper grip is substituted by a mushroom-shaped knob which is more ergonomically efficient than the traditional double grip.

Dividing – if the winch is without internal gearing – the length of the handle by the radius of the drum, you get the power ratio or mechanical advantage of the winch. If we have a 10-inch handle and a 2-inch radius drum, the power ratio will be 5. Be careful of where you measure the radius of the drum: the upper measure in the drawing below is the so-called 'neck' of the drum – where the turns of the sheet go – and is the correct one for calculating the power ratio, while the lower measure (not shown in illustration) is simply the diameter of the winch base and serves only to indicate how much room the winch will take up on deck. In the case of a winch with internal gearing, you get the final power ratio by also taking into account the transmission ratio, as we have seen above. Thus the higher the power ratio, the higher the reduction the winch applies to the force applied to it, and thus the less input force (that is, effort on your part) it needs.

A power monster is used for runners on America's Cup, where the Harken 990.3 str delivers a power ratio of 100 : 1. The power ratio indicated always refers to the slowest speed, which delivers the most power. In a two-speed winch this will be the second speed, in a three-speed winch the third and in a four-speed the fourth.

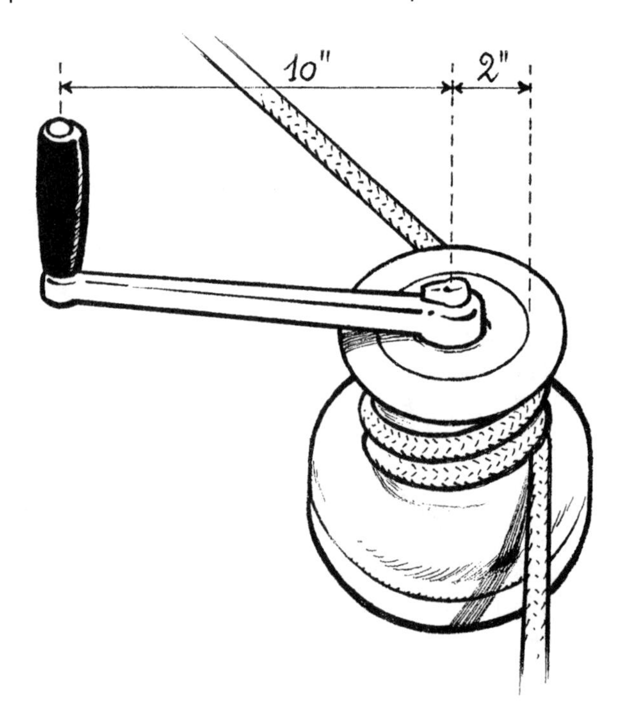

9.5

Measuring the power ratio of a winch: with a 10-inch handle and a drum radius of 2 inches, the ratio is 5.

The 990.3 str winch in the example has three speeds: in third the ratio[11] is 100 : 1, in second 25 : 1 and in first speed 2.5 : 1. This is quickly explained: third is used on the leeward side when sailing close-hauled to fine trim the genoa by a matter of millimetres, or if used for the runners to attain the tension required on the forestay. Second is used when completing a tack, particularly for bringing the last few metres of sheet into the boat as quickly as possible when the load is already high. First speed is used at the start of the tack when you need to haul in many metres of sheet with next to no load on it.

9.6
Carbon winch handle by *Lewmar.*

But let us look at the speed ratios in the various gears. In first, the speed ratio is 1. A winch with a 1 : 1 speed ratio in first gear is a winch that hauls in line very quickly, since every turn of the handle corresponds to a full turn of the drum, and this can allow many metres of sheet to be hauled in if the load is very low.[12] This is particularly useful these days since asymmetrical spinnakers, gennakers or code zeros have very long feet and thus very long sheets, which need to be hauled in quickly when gybing. In second, the ratio is 10 : 1 and in third, 40 : 1: this means that in second it takes 10 turns of the handle for just one turn of the drum, while in third it takes 40!

Table 9.1 Speed and power ratios of the 990.3 str winch

speed ratio			power ratio		
1	2	3	1	2	3
1:1	*9.9:1*	*40:1*	*2.5:1*	*24.9:1*	*100.27:1*

[11] With a 10-inch handle and the winch not powered by a coffee grinder.
[12] In first speed the power ratio is 2.5 : 1, so by applying 1 kg you only haul 2.5!

How can you determine the transmission ratio? It is easy. Take a 10-inch handle and, after marking the drum with a piece of tape, start turning the handle and count how many turns it takes to make the drum complete one revolution. This is the transmission ratio. Thus if two winches have drums of the same size, the one that needs fewer turns of the handle to make the drum complete a revolution is the faster.

But how can you work out whether one winch is faster than another in hauling in the sheet while tacking when you have one person turning the handle and the other hauling the genoa[13] sheet, if these winches have drums of different diameters? The following formula will tell you how many turns of the handle you need to haul in 1 m of line:

$$\textbf{Number of turns of handle} = \frac{1000}{\mathrm{DT} \times 3.14} \times \text{transmission ratio}$$

where 'DT' is the diameter of the neck of the winch drum (and the first speed gear ratio).

Types of winch

Winches were originally produced in the so-called *plain top* version: they finished with a cap that served as a cover and in which the seating for the handle was housed.

The *plain top* winch is very practical when we decide of our own free will always to hold the sheet in our hands, and paradoxically this means it is still popular today, seeing that all the primary[14] winches of America's Cup boats are of this kind. This is because the genoa trimmer[15] in the Cup needs to keep the sheet (or afterguy) under constant control. Apart from classic yachts and Cup boats, all others tend to use *self tailing* winches, which have toothed jaws that grip a certain length of the sheet while the winch is being turned and then release it. This has two advantages: on the one hand you can use a winch without necessarily needing someone to haul the line while you do so, and on the other once you

[13] The man who turns the winch is the 'grinder', the one who hauls the sheet is the 'tailer'.
[14] The genoa sheet winch is known as the primary winch; the others are the secondary ones, generally for spinnaker sheets or for general utilities, that is, spare winches without a precise function.
[15] The use of secondary winches for spinnaker sheets on Cup boats has long been abandoned, since the times of the international 12 m, in favour of a sophisticated and complex system on the primary winches. In fact, the primary winches are made with a sheave at their base that allows the spinnaker sheet to be held underneath the winch on which the afterguy is tensioned; after gibing the spinnaker sheet is moved on to the drum and the afterguy temporarily held by a stopper.

9.7

Harken 'plain top' winch on *Bona*, an international 8 m; right: line on a *self-tailing* winch (note the large number of turns taken before the line arrives at the jaws).

have finished trimming you can make the line fast in the *self tailing* jaws, which will thus do the job of a cleat.

A practical note. The jaws of a *self tailing* winch are designed to work at very low loads, so it is important to take as many turns as possible around the drum so as not to put on the jaws a load they were not designed to bear.

9.8
Harken 'top cleat' winch with a kind of cleat and a double opening in the head of the drum.

It is a good idea to have the arm of the *self tailing* mechanism in the 5 o'clock position, so that the line leaving it falls inside the cockpit. To establish this position, stand in front of the winch and rotate the arm of the *self tailing* (once you have freed it) to the position of the hour hand of a watch at 5 o'clock.

A half-way house between a *plain top* and a *self tailing* winch is the *top cleat*. On racing boats such as Kingfishers the primary winches are *self tailing*, since the long distances sailed do not require frenetically fast manoeuvres (and the crew is also small), but in other boats speed and agility in tacking are vital, and sometimes the *self tailing* is a hindrance. In fact, even a well-made and minimalist *self tailing* can foul the sheet with its protruding arm. But a winch with nothing protruding from its drum, as in the case of a *plain top*, does not create these problems, and if we add in a kind of cleat and a double opening on the top, we have a *top cleat* winch as illustrated below.

With a *top cleat*, as you trim the sheet you have to hold it in your hand (or pass it to a tailer) then, when you have finished trimming, make it fast to the cleat on top. With a *top cleat* you cannot turn the winch and recover the line at the same time, though when the sheet is cleated you can actually trim it just a little, about half a turn of the winch, but no more, because otherwise the line leaving the cleat would tend to foul itself under the turns of the sheet in tension on the drum.

9.9

Lewmar above deck winches (first two from left) and a Harken flush deck winch (third from left); below: drum support (right) and winch gearing (left).

A further classification of winches is into *flush deck* and *above deck* types. The former – of recent design – have a particular architecture in their gearing that allows them to have a very low base, and thus a relatively low angle of entry for the sheet, while the *above deck* winches are the traditional ones with a very tall base. *Underdeck* winches have their central axle protruding from the base with a special fitting for coupling to the winch to a system of coffee grinders or an electric or hydraulic motor.

9.10

Roller bearings (above, for lateral loads) and ball bearings (above right, for top loads) in a drum support; below: ring gear of a winch with its pawls.

Everything we have looked at so far concerns the exterior of the winch. But what is under the skirt? Under the drum, once you lift it off, you will find the drum support with the bearings on which the drum rotates. Underneath this are the gears that produce the transmission ratio. Above, the drum has its *self tailing* system or the *plain top* cap, while below it has the toothed wheel that meshes with the pinion gear to make the drum rotate.

The materials a winch is made from

There are basically two branches of winch: the cruising and the racing branches. Those who think that the latter are made without regard for reliability are quite wrong: the demarcation line between the two is weight and cost, not reliability. In fact, where in cruising winches the drum support is in bronze, in racing types it is in aluminium; where in the former the ball bearing races are in steel, in the latter they are in a special plastic material. Drums may be in bronze, aluminium or sometimes in carbon or Kevlar-carbon; and finally the gears, where possible, will like the bolts be in titanium instead of common steel.

9.11

Drum of a Harken 880.3 str winch in aluminium with self-tailing part in carbon.

Orientation of a winch

In reality there are pros and cons concerning the positioning of the winch on deck related to the efficiency of the winch, but we can in any case take this suggestion as valid while bearing in mind that the working load of a winch must be respected in all positions for the winch on deck. Because it is true that, even if we position the pinion at a tangent to the first turn on the winch, we do not always respect this condition in all situations. If, for example, the pinion of a primary winch is positioned correctly for the number three genoa sheet and the spinnaker afterguy (which arrive from forward of the winch) it may not be right for the code zero sheets, which arrive at the winch from astern!

Look at Fig. 9.14 (p. 169): here the turns on the drum of a winch are fouled because the angle of entry of the line is incorrect (the white line is almost horizontal with relation to the winch). The sheet should enter the winch at an angle of about 5° to 8° with respect to the base of the winch.

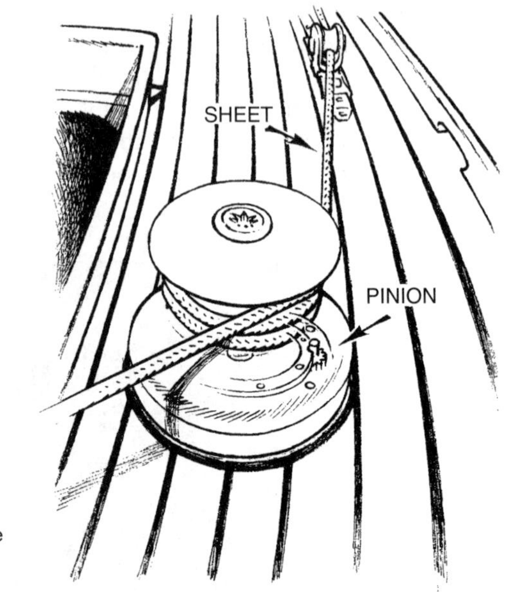

9.12

It is important that the pinion that meshes with the ring gear of the drum is at a tangent to the line formed by the sheet as it enters the winch.

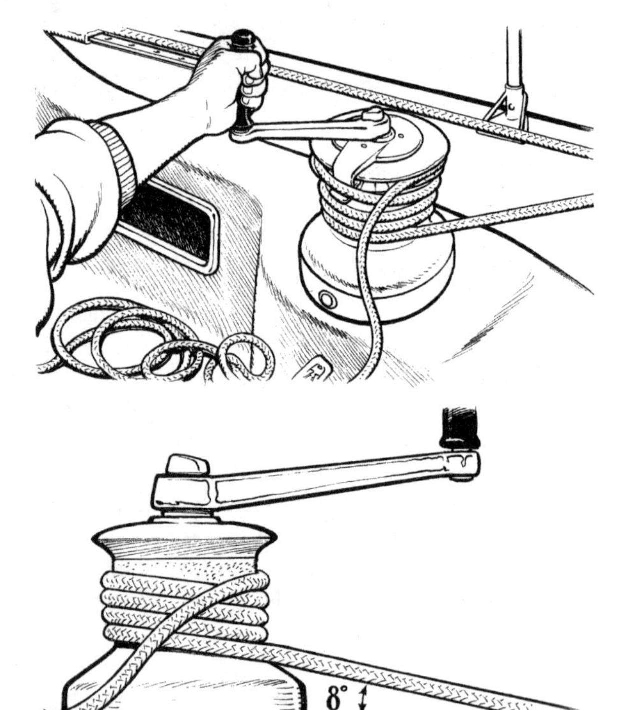

9.13

The correct angle of entry is between 5° and 8°; the positioning of the winch must take account of the position both of the pinion and of the first speed button, which must face in towards the cockpit.

Choosing a winch

How should we decide on a winch? First of all by comparison and through experience. If a 40 footer has always had a 48 winch and it has never given problems, you certainly cannot go wrong by staying with that size!

9.14

An overlapping turn on the winch drum caused by an incorrect angle of entry for the sheet.

Then a practical rule of thumb to decide quickly the size of a winch: on boats between 30 and 70 feet, add 13 to the length overall in feet and you will have the ideal power ratio for the primary winches.[16]

A more sophisticated method is to calculate the maximum working load on the sheet the winch will be used for and choose a winch with a suitable working load. The working loads of winches can be obtained from suppliers. Here too, as with purchases, pay attention to the difference between the maximum sustainable working load (that will not damage the winch while it's still and not in operation)[17] and the maximum operating working load (the most the winch can pull).[18] For example, a winch may be able to pull 2000 kg with no problems but also have a static working load of 4000 kg under which – thanks to such elements as a

[16] In the same way, halyard winches have a power ratio equal to the length overall of the boat in feet, while the secondary winches are one or two measures smaller than the primaries. This is a practical rule of thumb to give a rough idea, but always contact a trusted supplier before dimensioning deck plans.
[17] Static working load.
[18] Dynamic working load.

9.15
Barbarossa oil bath winch.

machined[19] support for the drum and the base – it may not work but will not suffer any permanent deformation.

Winch maintenance

A lot of mythology has been created around winch maintenance without coming to grips with the central issue, which is that it is not so much the grease or oil or the method of cleaning that are the key to the operation, but the fact that you carry it out at all! In fact, the basic problem of a winch is corrosion, which is caused by having a lot of different materials in contact with each other – what is more, in a marine environment.

Either you do as Barbarossa once did, which was to keep some models literally in an oil bath, or you need periodically to take your winches apart, put all the parts in a basin full of diesel oil, clean them with a brush, dry them with a rag that does not lose fluff, reassemble them and lubricate them with a thin film of grease spread on with a brush and not with your fingers (otherwise you will use too much), lubricating the pawls only with oil and never with grease. Use good-quality machine tool oil or that for weapons, and good-quality grease for marine use.

[19] This term is used when the part is not produced by moulding but cut out (on a lathe, with drills, etc.) from a solid block of material. This method of production costs much more than moulding, but ensures decidedly superior mechanical performance, eliminating the tiny bubbles typical of moulding processes that are liable to cause breakage.

Bibliography

Chapter 1

Jean-Marie Finot, *Elements de vitesse des coques*, Librairie Arthaud, Pairs, 1977.
Sam Svensson, *Manual pratico dei nodi marinareschi*, Mursia, Milan, 1979.
Antonio Santangelo, *Stabilitá e assetto della nave*, Patron, Bologna, 1970.
Gleistein Tauwerk, *Splice Book*, Geo. Gleistein + Sohn, Bremen, 2000.
C.A. Marchaj, *Sailing theory and practice*, Dodd, Mead & Co. Inc, New York, 1964.
Fabbroni & Vongher Yacht Riggers, *Bollettino Tecnico N° 1*, October 1989.
Marty Reick, Paterazzo sempre più potente, *Harken Technical Sailing*, n. 2 1993.
Marina Ropes Ltd, *Marina Ropes Catalogue*, UK, 1988.
Samson Yacht Ropes, *Samson Selection Guide*, USA, 1987; www.theamericangroup.com.
Brion Toss, *The Rigging Handbook*, Adlard Coles Nautical, London, 1998.
Funi d'acciaio, Teci company brochure, Milan, 1998.
Yale Catalog, *Ropes for Industry*, 1998; www.yalecordage.com.
New England Ropes: www.neropes.com.
Aramid Rigging: www.aramidrigging.com.
Navtec: www.navtec.com.
Artie Means, in 'Tech Review: Trick out your runners with high-tech line'. *Sailing World*, April 2001.

Chapter 2

www.loosco.com.
Douglas Phillips-Birt, *Sailing Yacht Design*, Adlard Coles Ltd, London, 1975.
www.futurefibres.co.uk.
www.gottifredimaffioli.com.
www.theamericangroup.com.

Chapter 3

The best of Sail trim, Granada Publishing Ltd, London; 1981.
Ed Baird, *Laser racing*, Fernhurst Books, Sussex, 1982.
Paul Elvstroem, *Elvstroem parla di sé e delle regate*, Mursia, Milan, 1971.

Chapter 4

Georges Devillers, *Manuale di arte marinaresca*, Mursia, Milan, 1982.
Brion Toss, *The rigging hand-book*, Adlard Coles Ltd., London, 1988.
Registro Aeronautico Italiano, *I cavi metallici: Ispezione e riparazione*, parte 1ª, 1990.
Loos Cable & Wire Company, www.loosco.com, USA.
Juan Baader, *Lo sport della vela*, Mursia, Milan, 1969.
Yves Louis Pinaud, *La practica della vela*, Mursia, Milan, 1979.
C.A. Marchaj, *Sailing Theory and Practice*, Dodd, Mead & Co. Inc., New York, 1964.

Chapter 5

Registro Aeronautico Italiano, *I cavi metallici: Ispezione e riparazione*, parte 1ª, 1990.
Loos Cable & Wire Company, www.loosco.com, USA.
Juan Baader, *Lo sport della vela*, Mursia, Milan, 1969.
Yves Louis Pinaud, *La pratica della vela*, Mursia, Milan, 1979.
C. A. Marchaj, *Sailing Theory and Practice*, Dodd, Mead & Co., Inc., New York, 1964.

Chapter 6

Roger Marshall, *Designed to Win*, Granada Publishing Ltd, London, 1979.
Matthew Sheahan, *Sailing rigs and spars*, Haynes, Someset, 1990.

Chapter 7

Wire Teknik, www.wireteknik.com.
Francis S. Kinney, *Skene's elements of yacht design*, Dodd, Mead & Co., Inc., New York, 1973.
W. E. Rossnagel, *Handbook of rigging: In construction and industrial operations*, Mc-Graw Hill, New York, 1975.
Hints and advice on rigging and tuning of your Selden mast Selden Co., Sweden, 2000.